Fostering Technology Absorption in
Southern African Enterprises

Fostering Technology Absorption in Southern African Enterprises

THE WORLD BANK
Washington, D.C.

ISBN: 978-0-8213-8818-1
eISBN: 978-0-8213-8886-0
DOI: 10.1596/978-0-8213-8818-1

Library of Congress Cataloging-in-Publication Data
Fostering technology absorption in southern African enterprises :a document of the World Bank.
   p. cm.
"This study was carried out by a team jointly led by Itzhak Goldberg (consultant) and Smita Kuriakose (economist, the World Bank) and comprised of David E. Kaplan (professor, University of Cape Town), Krista Tuomi (lecturer, American University), Reza Daniels (lecturer, University of Cape Town) and Peter Draper (senior research fellow, SAIIA)."
   Includes bibliographical references and index.
   ISBN 978-0-8213-8818-1 — ISBN 978-0-8213-8886-0 (electronic)
   1. Technology transfer—Africa, southern. 2. Technological innovations—Africa, southern. 3. Investments, Foreign—Africa, southern. 4. Africa, southern—Economic policy. I. Goldberg, Itzhak. II. Kuriakose, Smita. III. World Bank.
   HC900.Z9T44 2011
   338.968—dc23

                                                                                    2011022855

Cover photo: michaeljung/Shutterstock.com
Cover design: Quantum Think

Contents

Boxes

Figures

Tables

Acknowledgments

A team jointly led by Itzhak Goldberg (consultant) and Smita Kuriakose (economist, World Bank) and composed of David E. Kaplan (professor, University of Cape Town), Krista Tuomi (lecturer, American University), Reza Daniels (lecturer, University of Cape Town), and Peter Draper (senior research fellow, South African Institute of International Affairs) carried out this study. Justin Barnes (executive chairman, Benchmarking and Manufacturing Analysts SA [Pty] Ltd), Mike Morris (professor, University of Cape Town), and Eric Wood (professor, University of Cape Town) prepared substantive inputs, including background papers. The team wishes to thank the Human Sciences Research Council for providing access to the South Africa National Innovation Survey data. Chunlin Zhang (lead private sector development specialist, World Bank) and Smita Kuriakose prepared the final version of the book. The task was managed by current task team leader Chunlin Zhang and formerly by task team leaders Giuseppe Iarossi (senior economist, World Bank) and Antonio David (former economist at the World Bank), who provided overall guidance to and supervision of the team.

The team greatly benefited from constructive comments from peer reviewers Vandana Chandra (senior economist, World Bank), Mark Dutz (senior economist, World Bank), Shahid Yusuf (consultant, former adviser of the World Bank Institute), and John Gabriel Goddard (economist, World Bank). Valuable written comments were also received from the following colleagues in the World Bank: Shantayan Devarajan (chief economist, Africa Region), Alvaro Gonzalez (senior economist), Marco Scuriatti (senior operations officer), Irina Astrakhan (country program coordinator), Claus Pram Astrup (senior country officer), Sandeep Mahajan (lead economist), Tom Buckley (senior country officer), and Simi Siwisa (operation analyst). The team thanks Janice Tuten and Stuart Tucker, who provided support with the production of the book.

The team also gratefully acknowledges the extensive comments received and incorporated from the participants at the various consultation workshops conducted in all four countries. These workshops were organized jointly with the Graduate School of Business, University of Cape Town in South Africa on May 6, 2011; the Namibian Manufacturers Association on May 17, 2011; and the Lesotho National Development Council in Lesotho on May 19, 2011. A multistakeholder workshop was also held at the country office in Mauritius on May 12, 2011. All these workshops included representatives from government counterpart ministries, academia, research institutions, and the private sector. In addition to these stakeholder workshops, the team benefited from valuable comments from Garth Strachan (chief director, industrial policy) and Mlungisi Mthimunye (director, incentive development) of the Department of Trade and Industry, South Africa, and Julia Mungunda (acting deputy director for industrial development) and Nekuma Frans (head, technology research and development department) from the Ministry of Trade and Industry, Namibia.

The funding support of the Multi-Donor Trust Fund for Trade and Development is gratefully acknowledged. The work has been produced under the overall guidance of Ruth Kagia, World Bank country director for Botswana, Lesotho, Namibia, South Africa, and Swaziland; Marilou Uy, World Bank sector director for finance and private sector development, Africa Region; Gerardo Corrochano, former sector manager; and Michael Fuchs, acting sector manager, finance and private sector development, Africa Region.

About the Authors

Itzhak Goldberg is currently a visiting senior researcher at the Institute for Prospective Technological Studies of the European Commission's Joint Research Centre (JRC) and a fellow at CASE (Center for Social and Economic Research), Warsaw. He worked as an adviser for policy and strategy with the Europe and Central Asia Region of the World Bank from 1990 to 2009, where he was in charge of private sector development programs in various countries in the region and he was pivotal in the design and implementation of the privatization program of the government of Serbia. More recently, he has devoted his attentions to the economics of upgrading innovation and technology in the Europe and Central Asia Region and leading the series of studies on knowledge economy. He also published *Can Russia Compete?* with Raj Desai (Brookings Institution 2008). Prior to joining the World Bank, he was the chief economist and a member of the executive management board of Dead Sea Works Limited in Israel and adjunct associate professor of economics at Ben Gurion University in Israel. He also worked as a research fellow at the Hoover Institution in the United States in the late 1970s before he obtained his PhD from the University of Chicago in 1976.

David Kaplan is currently a professor of business government relations and professor of economics at the University of Cape Town, South Africa. He has long been involved in policy-oriented research and work with the government. Kaplan was the first chief economist of the Department of Trade and Industry for South Africa (2000–03) and has been engaged since then as a chief economist (part-time) at the Department of Economic Development and Tourism, Provincial Government of the Western Cape. He is currently working on a research project on developing local linkages to the mining sector in African countries.

Smita Kuriakose is an economist in the Finance and Private Sector Development Department in the Africa Region of the World Bank. Her expertise is in science, technology, innovation policy, and skills development issues with a view to promoting private sector development in Africa. In addition to this book, she has been a key team member in the ongoing skills and innovation studies in Mauritius and South Africa. She has also worked on lending operations in Ethiopia, Lesotho, and Mauritius. Previously, she worked on innovation and investment climate issues in the Europe and Central Asia Region, where she coauthored regional knowledge economy studies. Prior to joining the Bank in Washington, DC, she worked on the United Nations Link Project and in the Poverty Reduction and Economic Management Unit of the World Bank in India, where she focused on fiscal and macroeconomic policies. Kuriakose holds graduate degrees in economics from the University of Maryland, College Park and the Delhi School of Economics in India.

Krista Tuomi is an assistant professor in the international economic relations program at American University, Washington, DC. Prior to this post, she lectured at the University of Cape Town, South Africa, where she focused on growth theory, game theory, and macroeconomic policy. Tuomi also works as an independent policy analyst on southern Africa, advising on issues ranging from environmental cost-benefit analysis and governance capacity to growth constraints.

Chunlin Zhang is a lead private sector development specialist at the World Bank based in South Africa and a sector leader responsible for coordinating the Bank's operations in financial and private sector development in Botswana, Lesotho, Madagascar, Mauritius, Namibia, South

Africa, and Swaziland. Prior to his appointment to the South Africa country office in December 2009, Zhang worked in the Bank's Beijing office for 10 years. His analytical and policy advisory work covers such areas as state-owned enterprise reform, corporate governance, private sector development, public service reform, and technological innovation of enterprises. He also managed a World Bank technical assistance project that supports Chinese ministries and provincial governments in economic reform implementation.

Abbreviations

AGOA	African Growth and Opportunity Act
BEE	black economic empowerment
BPO	business process outsourcing
CEO	chief executive officer
CIS	Community Innovation Survey
CSIR	Council for Scientific and Industrial Research
DCCS	Duty Credit Certificate Scheme
EPZ	export processing zone
ES	Enterprise Survey
EU	European Union
GDP	gross domestic product
FDI	foreign direct investment
FETI	further education and training institution
HEI	higher education institution
HSRC	Human Sciences Research Council (South Africa)
ICT	information and communication technology
IPAP	Industrial Policy Action Plan
IPR	intellectual property rights
IT	information technology
ITC	International Trade Commission

MFA	Multifiber Agreement
MIDP	Motor Industry Development Programme
MNC	multinational corporation
MSC	Marine Stewardship Council
MUR	Mauritian rupee
NEET	not in employment, education, or training
NIS	National Innovation Survey
OECD	Organisation for Economic Co-operation and Development
OEM	original equipment manufacturer
R	South African rand
R&D	research and development
RFID	radio frequency identification
SAABC	South African Automotive Benchmarking Club
SACEEC	South African Capital Equipment Export Council
SADC	Southern African Development Community
SAIIA	South African Institute of International Affairs
SAR	special administrative region
SETA	Sector Education and Training Authority
SMEs	small and medium enterprises
SMMEs	small, medium, and micro enterprises
SPII	Support Programme for Industrial Innovation
Tekes	Finnish Funding Agency for Technology and Innovation
THRIP	Technology and Human Resources for Industry Programme (South Africa)
TVET	technical and vocational education and training
US$	U.S. dollar

Overview

Technological transfer and absorption play a critical role in development. As the Commission on Growth and Development pointed out, "learning something is easier than inventing it" (Commission on Growth and Development 2008, 2). Learning, while not without cost, allows technological latecomers to rapidly gain ground on the technological leaders by importing and absorbing technologies that are already created. "Knowledge acquired from the global economy is thus the fundamental basis of economic catch-up and sustained growth" (Commission on Growth and Development 2008, 41). Indeed, in all the cases of sustained, high growth (at a rate of 7 percent or higher for a period of 25 years or longer since 1950) that the commission examined,

> the economies have rapidly absorbed knowhow, technology, and, more generally, knowledge from the rest of the world. These economies did not have to originate much of this knowledge, but they did have to assimilate it at a tremendous pace. That we know. What we do not know—at least not as well as we would like—is precisely how they did it, and how policy makers can hurry the process along. This is an obvious priority for research. (Commission on Growth and Development 2008, 41)

This book seeks to understand how firms in southern Africa absorb technology and how policy makers can hurry the process along. It identifies channels of technology transfer and absorption through trade and foreign direct investment (FDI) and constraints to greater technology absorption, and it discusses policy options open to the government and the private sector in light of relevant international experience. The book is based on case studies of sectors and enterprises selected in four countries: Lesotho, Mauritius, Namibia, and South Africa.

The Relevance of Technology Absorption to Southern Africa

The relationship between technology absorption and catch-up growth is particularly relevant to southern Africa because those countries are facing tremendous competitiveness challenges and must rely on greater technology absorption to raise productivity and strengthen competitiveness to gain ground in the global market. An increased market share can then generate faster growth and create more jobs. Therefore, catch-up growth sustained by technological progress and productivity growth is the fundamental solution to unemployment and poverty alleviation.

Southern African manufacturing industries are squeezed by competition pressures on two fronts. On one front, countries in southern Africa face competition from knowledge-intensive economies in the industrialized countries, which occupied 70 percent of the global manufacturing market in 2005. On the second front, these southern African countries must compete with other developing and transition economies, including the emerging manufacturing giants in Asia, which derive much of their competitive strength from low labor costs. Together, other developing and transition economies enjoyed a market share of over 20 percent in 2005, leaving only 1 percent for Sub-Saharan African firms, of which South Africa alone accounts for more than half.

Southern African manufacturing industries may have very limited competitive advantage in terms of labor cost in competing with Asia. In particular, South Africa, the leading manufacturing economy in Africa, suffers a competitive disadvantage in terms of its labor costs. This general view was confirmed by a recent analysis of United Nations Industrial Development Organization data (Mengistae 2011) at the industry level. As illustrated in figure O.1, the advantage that South Africa used to enjoy in terms of lower unit labor cost over the other comparator countries has been lost as the other countries have been catching up since 2003. Of the six sectors studied, meat, fruit, and vegetable processing appears to be the only sector where South Africa has a clear advantage in lower costs.

Figure O.1 Unit Labor Costs in Selected Industries, 1996–2007

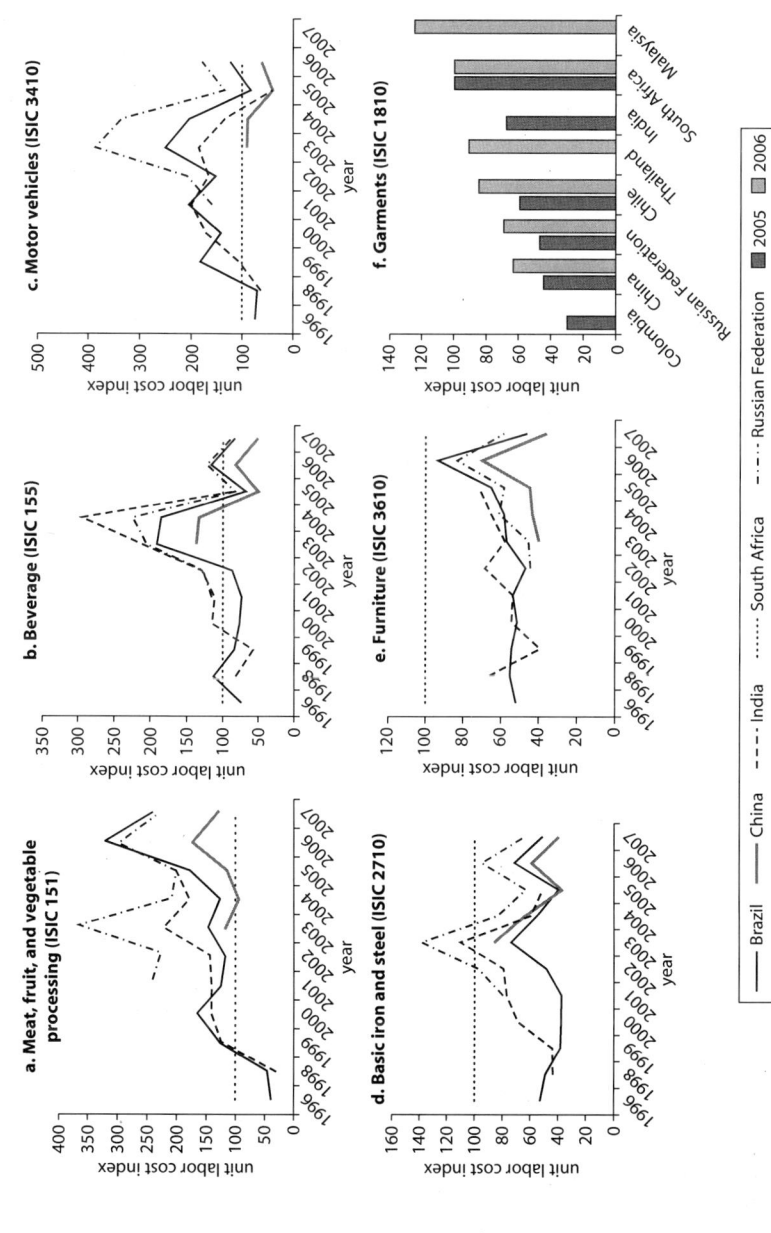

Source: Mengistae 2011.

Note: Index, South Africa = 100. ISIC = International Standard Industrial Classification.

3

To the extent that South African manufacturing industries are representative of manufacturing industries in southern Africa, the implication is that southern African economies must rely on increasing productivity to effectively compete with both knowledge-intensive and low-labor-cost economies. Southern African industries have significant potential to strengthen competitiveness through better management of costs of labor, infrastructure, and other inputs. However, in comparison with low-cost Asian emerging economies, southern Africa needs to gain more in terms of overall efficiency to compensate for its high labor and other costs. In other words, technological progress and total factor productivity are more important to Sub-Saharan Africa's manufacturing industries than they were to emerging Asian economies. However, efficiency and productivity do not appear to be the strengths of southern African economies. In South Africa, labor productivity growth, albeit strong, seems to have been driven predominantly by rising capital intensity. A close look at the export performance of the four southern African countries points to the limited technology content of their exports. In all four countries, exports are dominated by a combination of primary commodities and low-technology manufacturing products.

Channels of and Constraints to Technology Absorption

Southern African firms use multiple channels for technology absorption. For example, South African auto component firms entered technology agreements with global players to meet the demanding product standards required for export. Even after the global crisis in 2009, those who licensed technologies still spent 2.23 percent of their sales revenue on royalties. In Namibia, the meat-processing industry has made continuous efforts to upgrade technology, including the recent investment in radio frequency identification technology to trace cattle. In fish processing, companies use state-of-the-art production technologies, including electronic software to record and monitor production processes, intelligent portioning equipment, and sophisticated freezer systems. In the breweries sector, state-of-the-art technology is used at every stage of production and in the marketing and distribution processes.

Firms report acquisition of machinery and equipment, where advanced technology is embedded, as one of the leading channels of technology absorption. In the South Africa National Innovation Survey (NIS), 80 percent of the firms cite acquisition of machinery, equipment, and software as the primary channel by which they acquire new technology. In Mauritius, 54 percent of firms said the same in the World Bank Enterprise Survey.

Technology is also acquired from trading partners. In addition to technology agreements with global players, as in the case of the South African auto component industry, firms obtain technology and know-how from their suppliers of materials and equipment, foreign or domestic. Some equipment suppliers provide technical consulting services for installation, operation, and maintenance, which are valuable learning opportunities for the receiving firms.

Hiring skilled personnel from abroad has been important to Mauritian firms but not well used by other countries. In the textile industry in Mauritius, most critical high-level skills, including dye masters, designers, and market specialists, were imported from India. A significant proportion (35 percent to 40 percent) of the labor force in large textile and garment firms is acquired from abroad. Some seafood-processing firms hired a large number of expatriate workers to operate newly purchased equipment. The emerging information and communication technology sector is also aggressively recruiting expatriates because of a lack of locally available trained personnel in the field. In contrast, in other countries hiring expatriates remains a difficult process.

Skill shortages are cited as a constraint in most industries across all four studied countries. In the capital goods sector of South Africa, for example, firms reported a decline in the skills and competencies available. Skill shortages exist at the managerial, artisanal, and technical levels for positions such as welders and boilermakers. And this situation is being exacerbated by aggressive recruiting on the part of competitor firms, especially in Australia, which have been very successful at recruiting skilled South Africans. By contrast, South African firms complain that they have major problems in securing the necessary work permits for expatriate labor. Consequently, the shortage of skills is threatening South Africa's global technological leadership position in mining and mining-related activities, which is one of the few fields where it has technological strength in the global market.

The role of FDI as a channel for technology absorption varies greatly. Whereas Mauritius's success in the textile and garment industry is rooted in the FDI it attracted to this industry in the 1970s, the success of Lesotho in attracting FDI in the same sector in the recent decade has generated only export growth without any significant spillover effects into the rest of the domestic economy.

Southern African firms' research and development (R&D) activities appear to be limited, and industry-research collaboration is weak, which may have restricted firms' absorptive capacity. In the chemical industry in South Africa, for example, other than the dominant firm Sasol, other firms conduct very little R&D. Collaboration between firms and R&D institutions, including tertiary education institutions and publicly funded scientific institutes, appears to be very limited in South Africa. In the NIS, only 1.8 percent of the 600 firms surveyed rated their links with such institutions as important to their innovation activities. In the interviews, a similar picture emerged. Very few firms had any engagement with these institutions, and where such an engagement occurred, it tended to be quite limited. Although having limited industry-research links is not uncommon in other countries or on a regional level, the figures for South Africa still represent a rather low level of collaboration. For example, the Community Innovation Survey data show that 8 percent of the firms in the EU-27 countries have links with universities and technikons, and 5 percent with government research institutes.

Policy Options for Greater Technology Absorption

Governments can support innovation in general and technology absorption in particular in a variety of ways (see box O.1 for the experience of countries of the Organisation for Economic Co-operation and Development [OECD]). At the most basic level, effective government policies can create an institutional base that establishes openness to trade, improves the business environment for domestic and foreign investment, establishes effective intellectual property rights regimes, and enhances knowledge flows and learning. When the business environment is supportive, firms driven by profit motives will seek to update their technology in the best way they can. Beyond those general policies, however, governments may need to intervene at the industry and firm levels to directly address market failures.

Given the channels of and constraints to technology absorption in southern African firms, public policy should first focus on getting the basics right through fostering entrepreneurship, improving the investment climate, and strengthening competition. Second, more focused policy actions can be taken in four particular areas: learning through trade, FDI spillovers, skill development, and R&D activities.

Getting the Basics Right

Constraints to technology absorption in southern African firms do not differ very much from those to economic growth in general. Before taking more proactive actions to correct market failures, governments in these countries are well advised to get the basics right.

Box O.1

A Brief History of Innovation Policy in OECD Countries

In the first part of the 20th century, innovation policy as such did not truly exist. It was gradually developed as a way to promote the industrial competitiveness and social welfare of countries, as a complement to actions taken by governments to develop defense technologies, as initiated in World War II. Innovation policy has emerged gradually as a policy distinct from both science and industry policies. The evolution of government efforts to encourage innovation over the second part of the 20th century can be summarized as follows.

1950s. This decade saw the building of modern science systems in the industrialized world. In some countries, piecemeal measures were occasionally adopted to reduce identified weaknesses in the innovation process, including the creation of the National Research and Development Corporation in the United Kingdom (1949), the aim of which was to facilitate the promotion and diffusion of inventions from public laboratories and universities. Among others, France established sector-specific technical centers to help industries with technical research assistance and information, and Germany set up the Fraunhofer system of applied research and development (R&D).

1960s. Two trends were noticeable. First was the launch of large-scale programs in strategic areas such as space, nuclear technology, and oceanography, in countries such as France, the United Kingdom, and the United States. Second was the emergence of the concept of innovation policy distinct from science policy. The seminal report in this area is the Charpie Report, published in the United States at the request of the Department of Commerce in 1967 (U.S. Department of Commerce 1967). It stated clearly the need to act on diverse factors affecting the innovation climate, such as university-industry relations, venture capital, procurement policies, tax incentives, and competition laws—with particular attention to small enterprises and individual inventors, which it presented as the main source of innovation.

1970s. The decade saw a proliferation of government measures to promote innovation in the form of civilian technology programs, R&D incentive schemes for in-house efforts in the business sector, and university-industry collaboration. This was particularly evident in Europe and Japan, which were trying to decrease the increasing technology gap with the United States. The oil crisis of 1973–75, and the subsequent economic slowdown, also led to renewed interest in innovation policies. A concerted effort was made to develop an institutional framework

(continued next page)

Box O.1 *(continued)*

for innovation. For instance, in the United States, the Small Business Administration was strongly involved in support for small firms, which were perceived as a key source of innovation (as in the Charpie Report); the National Science Foundation, for its part, supported basic research; and the various sector agencies (defense, commerce, interior, and so on) all had technology-related programs.

1980s. Two major trends emerged. First was the development of regional technology and innovation policies, owing to an increased perception that innovation flourishes in sites with a concentration of talent, knowledge, and resources. Therefore, building critical mass was considered important, and major programs were set up to build science parks or "technopolises" (in Japan, for example). The need was also felt to act as closely as possible to entrepreneurs and potential innovators in order to help them more efficiently. Hence, territorially decentralized innovation policy initiatives proliferated, often encouraged by central governments through various schemes (such as decentralized antennas of central innovation agencies, or matching funds provided to local governments). The second major feature of this decade was the emergence of the notion of national innovation systems, which emphasized the interactions among key actors and communities (research, business, education) as a source of the innovative dynamism of countries and the need for governments to strengthen such systems through appropriate policy actions.

1990s. Inspired by the concept of national innovation systems as well as the acceleration of the globalization process, the spread of information and telecommunication technologies, and the emergence of new technologies such as biotechnologies, governments systematically engaged in building innovation policies that encompassed established policy fields. In the traditional science policy field, efforts were made to connect basic research more closely to applications. In industry policy, horizontal actions to boost innovation efforts were perceived as an efficient way to replace traditional policies to "pick winners," which were criticized for their inefficiency and ideological inadequacy. The Nordic countries have probably been the most active in adopting this approach. In the mid-1990s, Finland, for example, created two key institutions for promoting innovation: Tekes, the technology agency in charge of supporting innovation directly with a very significant budget; and the Science and Technology Policy Council, chaired by the prime minister, with the active participation of all ministers (including finance), which seeks to improve the innovation climate in all relevant policy fields and is directly inspired by the concept of the innovation system.

(continued next page)

Box O.1 *(continued)*

2000s. The notion of innovation policy has become very fashionable, and many countries have adopted it, as evidenced by the development and proliferation of OECD innovation system and policy reviews. Initially pioneered in the mid-1980s, such reviews now respond to strong demand, not only from the "old" OECD members, but also from the transition economies that have recently joined the OECD, as well as dynamic emerging economies from different parts of the world, such as Chile and China.

Source: World Bank 2010a, 56–57.

Creating an environment conducive for entrepreneurs to create wealth through innovation is critical to fostering new ventures and creating jobs. Government support can include a business environment that allows failure and company exit as a necessary part of entrepreneurial learning, company incentives that favor entrepreneurs with good ideas, instruments that enable entrepreneurs to access capital for start-ups, and flexible labor market policies that enable firms to expand by attracting the most skilled and talented workers from outside the firm or the country.

Greater competition is particularly critical for technology innovation and absorption in South Africa. In the NIS, of the constraints listed, the greatest number of firms (26 percent) cited the market being dominated by established firms as a reason for not innovating. In one of the firms interviewed in the book, the management indicated clearly that the competitiveness of its products depends not so much on innovation and technology upgrading as it does on the price of one input, which accounts for 75 percent of the total cost of the product and is supplied by one dominating domestic firm. This observation adds to one of the key findings of the recently completed second Investment Climate Assessment of South Africa, namely, the need for a more activist and innovative competition policy aimed at tackling barriers to entry and innovation. A widely shared view is that industrial concentration in South Africa has hindered market competition, with adverse implications for productivity and employment in all sectors (Government of South Africa 2010). In addition to competition policy, trade and investment policies should be used to generate positive change in market structure and introduce competition pressure from abroad.

A concerted government effort is also needed to improve the investment climate in southern African countries. A vast majority of the firms interviewed were constrained by logistical and infrastructural factors. High costs and lack of reliable rail transportation, harbor access, and power provision skew technological choices by favoring road transport, or the installation of company-owned power generators, for example, adversely affecting exports. Although some South African firms are efficient and use technology well in house, they face key constraints in the broader environment within which they operate. Improving the investment climate is particularly important for technology absorption through FDI. First, it helps increase the inflow of FDI. Second, by strengthening the attractiveness of the host economy, a better investment climate allows the host country to negotiate more easily with foreign investors to increase technology and skill transfer.

Skill Development

Skill supply in South Africa has failed to respond adequately to demand. Despite a persistent shortage of skills and the premium attached to skills by the labor market, the technical and vocational education and training (TVET) system and the higher education system failed to increase the supply of skilled labor to keep pace with the evolving demand. In 2007, 2.8 million youth in the age cohort of 18–24 years were not in employment, education, or training (known as "NEETs"). Total enrollment of public higher education institutions (HEIs) and further education and training institutions (FETIs) was only 1.1 million. Enrollments in apprenticeships and artisan-related learnerships reached only 0.03 million in 2009–10. The growth of capacity and output of the postsecondary education system does not seem to have responded well to the persistent skill shortage. Enrollment in public HEIs increased by only 0.23 million in 1999–2009, while the total number of learners in public FETIs increased by 0.15 million. These low enrollment figures are compounded by strikingly low completion rates. In 2009, the public postsecondary education system added 40,973 science, engineering, and technology graduates; 33,788 business management graduates; 8,112 master's degree holders; 1,380 PhD degree holders; and an unknown number of FETI graduates to the labor force. In the meantime, the working-age population increased by 0.8 million, from 30.9 million to 31.7 million, while the size of the labor force was 17.5 million.

The persistence of skill shortages has its roots in both skill demand and supply but is mainly a result of multiple constraints on the supply side.

South Africa's economic growth since the mid-1990s has generated increasing skill intensity in labor demand, which obviously contributed to the skill shortage. However, this situation has led to a significant premium for skills in the labor market, as indicated by the growing disparity between the wages of skilled and unskilled labor. A well-functioning post-secondary education system would have responded with a dramatic expansion of investment in capacity and an increase in supply. That this has not happened suggests the existence of serious constraints on the supply side. The first constraint is the well-documented inadequate quality of school education, which has resulted in uneven preparedness among entrants to public HEIs and FETIs. Known as the *articulation gap*, a mismatch results between the entry assumptions of those entering higher education and the actual competencies gained by graduates. The inability to attract and retain sufficient qualified academic staff is another constraint. Financial constraints, inadequacy of infrastructure, poorly developed links with business and industry, and internal governance weaknesses of public institutions may have all been responsible for the dismal supply-side response to skill demand.

South Africa needs to take urgent action on a very large scale to dramatically increase the supply of skilled labor. The need for large-scale action is justified by the magnitude of the challenge. The total number of NEETs and new entrants to the labor market, which could potentially be in the range of 3 million to 4 million people, renders any marginal improvement in the postsecondary education system, with a capacity of about 1.3 million, insignificant. This urgent action is likely to entail joint efforts from both the public and the private sectors. A reform strategy oriented toward a public-private partnership, whereby the government concentrates on financing and quality assurance, mobilizing both public and private service providers to increase service provision, may have a better chance of success.

First, in view of the severe social consequences of the large number of NEETs, the government should consider extraordinary short-term actions to provide some form of learning opportunities to NEETs if formal education, employment, and training are not possible. This intervention could serve as a temporary measure for addressing the poor quality of school education[1] as well as alleviating the social pressure of youth unemployment. Although the capacity of the higher education sector also requires dramatic expansion, especially in the science, engineering, and technology fields, priority should be given to the TVET sector and, in particular, workplace learning. Conceivably, many more postschool youth would find entering the labor market easier either (a) through the FETI route or (b) by participating in apprenticeships or learnerships than by using the higher education route. Workplace learning is thus critical to their success, and its absence is a major weakness of the existing TVET system. Finally, South Africa needs to strengthen its position in the international competition for talent. In addition to a sound financing regime, the recruitment of qualified educators is critical to capacity expansion. Because people with high-end skills are characterized by greater international mobility, South Africa will need to participate in the international competition for talents more effectively to reduce "brain drain" and maximize "brain gain." Concerted efforts are needed to sufficiently strengthen South Africa's position and attractiveness in this competition.

Skill mismatch is an issue in Mauritius, Namibia, and Lesotho as well. Similar to the case of South Africa, student attrition of 31 percent in Mauritius is a major concern, and only 13.5 percent of total primary school entrants obtain their school certificate, which is very low compared to 90 percent of students graduating from primary to secondary education in Mexico, Turkey, and Vietnam (World Bank 2011b). This high attrition rate after primary education contributes to a low skill base in Mauritius, with students as young as 12 years of age who are not successful in passing the certificate of primary education after two attempts being forced to attend prevocational training outside the general education stream. The prevocational stream is not geared toward teaching the student core literacy skills, making it extremely hard for students who are in the prevocational stream to pursue higher studies or integrate into the labor market. Mauritius is also facing a skill shortage in the teaching community. Although the percentage of PhD holders among the academic staff increased to 40 percent in 2007–08, that figure is still low when compared to international standards. Modest salaries paid to academic staff hinder attracting highly skilled Mauritians or foreign academics to join the universities. Furthermore, the rapid growth of the TVET system and a lack of coordination between the private providers have led to fragmentation and a certain amount of duplication. Although unlike the case in South Africa, a high level of dialogue occurs between the providers and the private sector, nevertheless the current TVET programs are not designed to meet the demands of the labor market. Namibia and Lesotho do not have an adequate number of skilled professionals graduating from the universities, and a large proportion of graduates are unequipped with the skills required by the private sector.

Mauritius needs to introduce measures to increase the scale and quality of its workforce. Changes should be made in the education system, giving growing importance to science, technology, and mathematics. Given that Mauritius aims to move toward a knowledge-based economy, demanding new higher-end skills, a fundamental restructuring of the current education system is called for to meet the skill demand. On the supply side, investing in science and engineering education to strengthen Mauritius's technical workforce should be a government priority. Moreover, the age of students who get into prevocational training could be increased, and they should be equipped with the requisite learning skills for the labor market. The TVET system should be realigned to cater to the changing demands of the economy, with greater emphasis on new sectors such as information and communication technology, which the current system does not address, and to provide greater numbers of technicians and more professional skills. Reforms also are needed in the training delivery sector, including provision of adequately trained instructors and better-equipped facilities. In the higher education sector, similar reforms would include curriculum revisions to better align with the needs of emerging industries and to improve the quality of academic staff. Because those reforms would have implications for public debt, the introduction of income-contingent loans, whereby students pay back loans in relation to the income they derive, could be explored.

Addressing the labor market constraints should be a priority in Namibia and Lesotho as well. The ease of importing skilled labor into Mauritius stands in stark contrast to the situation in South Africa, or in Namibia where hiring expatriate labor and obtaining the requisite work permits are extremely difficult. A short-term solution for partially alleviating the severe skill shortage in South Africa and Namibia would be to facilitate the hiring of skilled personnel from abroad. The possibility of relaxing the black economic empowerment rules while skills are scarce should also be considered. In Namibia, expediting the process of obtaining a work permit and granting permits for longer than the current three months should be considered. Labor costs are substantially higher in Lesotho than in comparable Asian countries (R 890 a month versus R 250 to R 450 per month), especially given the low skill level and high turnover rates. The skill base of the Lesotho labor pool should be expanded. Though the recent establishment of a basic training facility for machinists is recognized as a positive development by the firms surveyed, significantly more needs to be done to bolster the skill profile of the workforce—at both machinist and more technical levels.

Other alternatives include providing greater links with the University of Lesotho and addressing the prohibitive costs associated with bringing technical training experts to Lesotho. Although Mauritius is able to alleviate some of its skill shortage by hiring expatriate labor, given the perceived skill shortages, Mauritius still needs to introduce measures to increase the scale and quality of its workforce. Collaborating with overseas universities (for example, in the United Kingdom and Australia) could be an additional channel for effective human resource development. Foreign universities can help develop and upgrade curricula and teaching materials as well as provide teaching staff. These collaborative initiatives could be undertaken by tapping into the Mauritian diaspora. Ongoing efforts initiated by the government of Mauritius could be further strengthened to attract skilled Mauritian researchers and academics to contribute to and encourage collaborative programs between Mauritian nationals abroad and research institutes and universities in Mauritius. These efforts could include taking Mauritian students into their labs or research institutes and providing them with lectures when they return home to visit their families. The National Youth Council in Taiwan, China, is an example of an effective program that garnered synergies by connecting local businesses with skilled migrants abroad.

Learning through Trade
Given the importance of trade, especially the acquisition of machinery and equipment, as a channel of technology absorption, governments must facilitate learning through this channel. In addition to ensuring general openness to trade and a supportive business environment, case studies of southern African firms and experience in other developing economies point to the need for greater government support for learning through trade, including learning from exporting, importing, and knowledge transfer supported by what is known as "technology diplomacy," in which governments make use of their bargaining power in trade to promote technology transfer to their domestic economies. One instrument can be a matching grant scheme designed to defray part of the cost incurred by firms, especially small and medium enterprises, engaging in learning through trade.

FDI Spillovers
Although attracting FDI remains a priority for southern African economies, more proactive actions are needed to increase the benefits of spillovers from FDI. Existing empirical evidence shows that the theoretically postulated spillover effects do not materialize automatically just because

a country is able to attract FDI in the first place, as confirmed by the case of Lesotho's textile industry in comparison with, for example, the case of Bangladesh. The first-order challenge, nonetheless, remains to attract more FDI. South Africa, in particular, seems to have lagged behind its Asian peers in attracting FDI into manufacturing industries (World Bank 2010b). This necessity makes maximizing the spillover benefits from existing FDI even more imperative.

Proactive actions in technology absorption could be useful to southern African countries. Three sets of such actions can be considered. The first is providing incentives to foreign investors to motivate them to engage in deliberate actions of technology transfer to the local economy. Such incentives should be tied to ensuring the performance of these investors' results in technology and skill transfer and could be provided by indigenous private firms with appropriate government support. Although this effort may raise understandable concern when countries are struggling to attract more FDI, these incentives need to be designed to fully compensate the extra cost that FDI firms would have to incur in engaging in technology and skill transfer. The second action includes incentives to encourage domestic firms' learning efforts. The third may involve measures and actions to be taken to strengthen absorptive capacity. In addition to skill development, which is an obvious priority, the development of basic infrastructure for technological progress, such as the system of metrology, standardization, testing, and quality, as well as technical advisory services, is another potential priority.

R&D Activities

Support to R&D and research-industry collaboration is of strategic importance in building absorptive capacity and ensuring research is demand driven. South Africa would be well advised to consider restructuring the R&D tax incentive to make it easier to access, particularly for small firms; extending the list of qualifying expenditures to include more applied R&D, as appropriate for supporting technology absorption; and allowing a carryforward of the tax deduction to provide an incentive for R&D activities whose returns do not materialize within one year. In South Africa, an OECD study found the Technology and Human Resources for Industry Programme (THRIP) managed by the Department of Trade and Industry to have been very effective in integrating the development of research-capable human resources with industry-university cooperation in R&D, and the program has been recognized internationally as particularly successful when compared with similar schemes in other countries. In fiscal 2008/09, 207 small, medium, and micro enterprises and 106 larger firms were engaged in THRIP projects. This success should be scaled up. A study could be undertaken to assess how THRIP might be extended more widely. Such a study might focus on why firms located in sectors in which collaboration would be expected do not engage with THRIP. Other mechanisms can also be considered (see table O.1).

Intracountry technology diffusion also deserves government support. Although technology absorption from the global technology pool is often emphasized under the assumption that advanced technologies are mostly developed in industrialized countries, this focus is by no means the entire story of technology absorption. Technology transfer and absorption can take place within one country, one industry, or even one firm. In South Africa, this observation applies particularly to mining and mining-related equipment and specialist services, where a major cluster of South African firms has significant expertise along the global technology frontier. The challenge is to support the spread of these technologies and companies into new non-mining-related products and markets, which could be considered as a priority area for support in various government programs, such as the Industrial Policy Action Plan 2 and THRIP.

The subject of technology absorption in southern Africa requires further research. The scope of this book is limited and by no means exhaustive. Case studies have serious limitations when insights gained are generalized. A carefully designed survey would be a desirable direction for future research. Given the importance attached by firms to the acquisition of machinery and equipment, more effort should be devoted to looking into how technologies are adopted, assimilated, and absorbed through the importation of machinery and equipment. The role of the private sector in skill development deserves special investigation when more data are available. R&D activities should differ when they are oriented to technology absorption rather than technology creation. How they should differ in the context of southern African manufacturing industries is another potential area on which to focus. Finally, the overall policy direction—and hence options for action—is a subject that requires a great deal of research.

Because countries, industries, and firms vary dramatically in many dimensions, including the nature of the technology in question, wherever a recommendation from this book is found to be worth trying, action plans must be tailored to the specific local conditions. In this regard, recommendations in this book are more about directions and principles for further policy discussion than prescriptions.

Table O.1   Direct Instruments for Supporting Business R&D

Instrument	Advantages	Disadvantages
Tax incentives for R&D	Provides functional intervention, not picking of winners Offers less distortion, more automatic Generally requires less bureaucracy to implement, although advisable to have monitoring and spot checks	Has unclear fiscal costs in advance, which could be high Is difficult to ensure that R&D increase is induced by tax incentives (additionality) Is not very relevant for start-up firms that do not yet have taxable revenue streams Is a blunt instrument; cannot target specific companies, although it can target specific sectors
Grants for R&D projects	Allows specific targeting on case-by-case basis Can control amount of subsidy granted Can be given in tranches against defined goals Can be structured as matching grants that may help improve quality or efficiency	Requires large bureaucracy to administer May not select the best project Is also difficult to ensure additionality
Accelerated depreciation for R&D equipment	Reduces the capital costs of R&D projects	Does not provide incentives for noncapital costs such as personnel and material inputs
Duty exemptions on imported input into R&D	Reduces cost of world-class inputs if country otherwise has high import duties	Results in loss of tariff revenue; is distortionary to the extent that it favors R&D over other activities
Venture capital to facilitate commercializable research results	Helps overcome financial market failure in making capital available to start-ups with no collateral or track record	Requires detailed knowledge of sectors to evaluate technical and commercial prospects Is often not successful because of limited deal flow and shortage of technoentrepreneurs Also requires developed stock markets so investors can sell off shares and reinvest in new projects

Source: World Bank 2010a, table 5.9.

Note

1. The matriculation completion rate among black and colored youth is less than 40 percent. Because matriculation is the lowest recognized qualification in South Africa, a large segment of the young population is left with nothing to signal its ability on the labor market (World Bank 2011a).

References

Commission on Growth and Development. 2008. *The Growth Report: Strategies for Sustained Growth and Inclusive Development.* Washington, DC: World Bank on behalf of the Commission on Growth and Development.

Government of South Africa. 2010. "Cabinet Statement on the New Growth Path." Government Communications (GCIS), October 26, 2010, South Africa. http://www.info.gov.za/speech/DynamicAction?pageid=461&sid=14034&tid=23221.

Mengistae, Taye. 2011. "Are South African Wages Too High or Growing Too Fast?: A Comparison of Manufacturing Pay and Productivity in Selected Middle Income Economies" (Draft). Background paper for ongoing study on economic diversification in South Africa, World Bank, Washington, DC.

U.S. Department of Commerce. 1967. *Technological Innovation: Its Environment and Management.* Washington, DC: U.S. Government Printing Office.

World Bank. 2010a. *Innovation Policy: A Guide for Developing Countries.* Washington, DC: World Bank.

———. 2010b. *South Africa: Second Investment Climate Assessment: Business Environment Issues in Shared Growth.* 2 vols. Washington, DC: World Bank.

———. 2011a. "Closing the Skills and Technology Gaps in South Africa" (Draft May). World Bank, Washington, DC.

———. 2011b. "Skills and Technology Absorption in Mauritius" (Draft May). World Bank, Washington, DC.

CHAPTER 1

Technology Absorption and Its Determinants

An Introduction

Smita Kuriakose, Chunlin Zhang, and Itzhak Goldberg

Technological transfer and absorption play a critical role in development. Although economic theory considers technological progress to be a key factor for sustained long-term economic growth and job creation, technology absorption is particularly a driver for "catch-up growth." As the Commission on Growth and Development pointed out, "learning something is easier than inventing it" (Commission on Growth and Development 2008, 2). Learning, while not without cost, allows technological latecomers to rapidly gain ground on the technological leaders by importing and absorbing technologies that are already created. "Knowledge acquired from the global economy is thus the fundamental basis of economic catch-up and sustained growth" (Commission on Growth and Development 2008, 41). Indeed, in all the cases of sustained, high growth (at a rate of 7 percent or higher for a period of 25 years or longer since 1950) that the commission examined,

The authors gratefully acknowledge background papers on trade and foreign direct investment trends by Peter Draper and Krista Tuomi, respectively.

the economies have rapidly absorbed knowhow, technology, and, more generally, knowledge from the rest of the world. These economies did not have to originate much of this knowledge, but they did have to assimilate it at a tremendous pace. That we know. What we do not know—at least not as well as we would like—is precisely how they did it, and how policy makers can hurry the process along. This is an obvious priority for research. (Commission on Growth and Development 2008, 41)

This book seeks to understand how firms in southern Africa absorb technology and how policy makers in southern African countries can hurry the process along. It identifies channels of technology transfer and absorption through trade and foreign direct investment (FDI) and constraints to greater technology absorption, and it discusses policy options open to the government and the private sector in light of relevant international experience. The book is based on case studies of sectors and enterprises selected in four countries: Lesotho, Mauritius, Namibia, and South Africa. It is designed to support a number of the strategic priorities of the respective governments, which have all highlighted the need for progress in economic diversification and competitiveness with a view to increasing job creation.

As such, this book complements a number of other ongoing World Bank studies. In the case of South Africa, a parallel study titled "Closing the Skills and Technology Gaps in South Africa" (World Bank 2011a) looks at technology innovation closer to the global frontier taking place in certain key industries in tandem with skill development. In the case of Mauritius, this book complements and builds on the latest Investment Climate Assessment and an analytical work on innovation and skills being undertaken in a forthcoming study titled "Skills and Technology Absorption in Mauritius" (World Bank 2011b). The World Bank has also completed a regional Investment Climate Assessment for the Southern Africa Development Community that focuses on the institutional and regulatory framework in which firms and markets function to increase cross-border harmonization.

Technology absorption is differentiated from innovation in this book as follows. *Innovation* is understood as the development and commercialization of new, unproven technologies and untested processes and products. *Technology absorption*, in contrast, is defined as the application of existing technologies, processes, and products that have been proved and tested in new markets where commercial applications are not fully known. Simply put, innovation involves creation of "new-to-the-world" technology, whereas absorption is about "new-to-the-firm" technology. This distinction does not, however, preclude important complementarities between innovation and absorption. The ability of an economy to research and develop new technologies increases its ability to understand and apply existing technologies. Similarly, the absorption of cutting-edge technology inspires new ideas and innovations.

This book has three chapters. This introductory chapter discusses the relationship between technology absorption and growth and its relevance to southern Africa. That discussion is followed by a brief conceptual framework of the determinants of technology absorption. The chapter ends with an account of the empirical methodology of this book. Chapter 2 analyzes the specific channels of technology absorption in southern African enterprises and the key constraints that those enterprises face when trying to effectively invest in technology absorption. Policy options open to both the government and the private sector are assessed in chapter 3. Supplementary technical details are provided in the annexes to the respective chapters.

Technology Absorption and Catch-up Growth

Given the critical role of technology absorption as identified by the Commission on Growth and Development, clearly no developing economy can afford to lag behind in technology absorption. As succinctly argued by Lin and Monga (2010, 7), "Compared with developed countries whose industries are located on the global frontier and their industrial upgrading and diversification rely on their own generation of new knowledge through the process of trial and error, developing countries in the catch-up process can move within the global industrial frontier and have an advantage of backwardness. That is, developing countries can rely on borrowing the existing technology and industrial ideas from the advanced countries."[1]

Technology absorption drives catch-up growth through a complex process in which enhancement in productivity and hence increased competitiveness are crucial. As illustrated in figure 1.1, in a conducive investment climate, technology absorption raises the efficiency of the use of labor and capital, resulting in higher productivity and accelerated growth. In addition to capital investment, this requires the inflow of both codified and tacit knowledge, for which trade and FDI are two key channels, as

Figure 1.1   Technology Absorption and Economic Growth

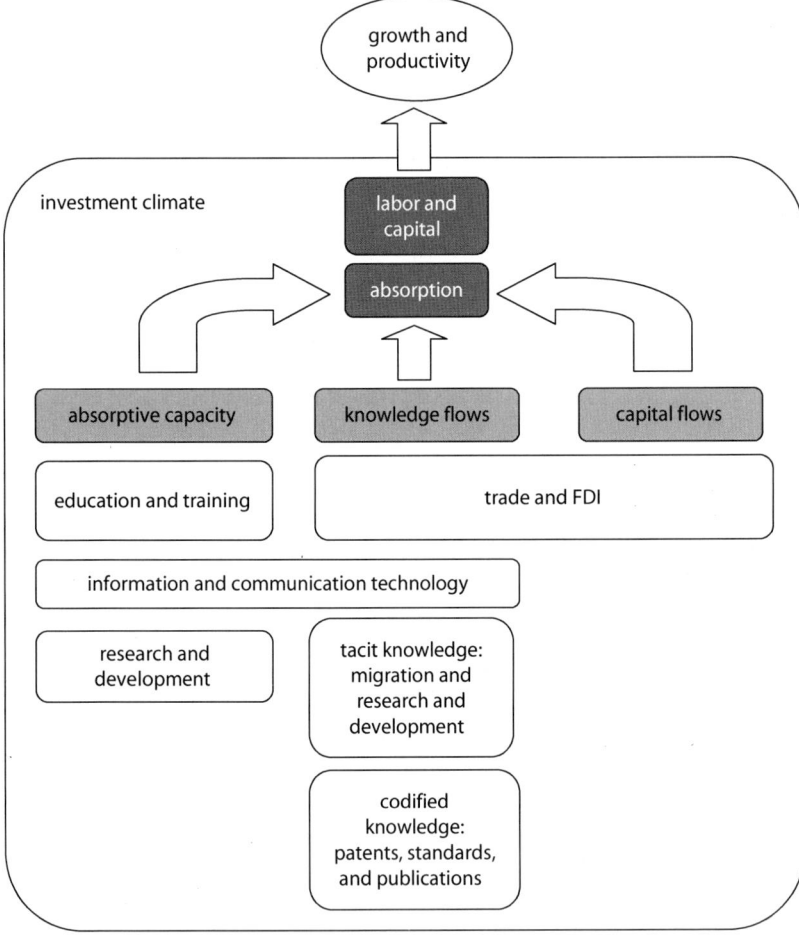

Source: Goldberg and others 2008.

well as an adequate capacity of the economy to absorb the knowledge it has access to.

The Relevance of Technology Absorption to Southern Africa

The general relationship between technology absorption and catch-up growth is particularly relevant to southern Africa because those countries are facing tremendous competitiveness challenges and must rely on greater technology absorption to raise productivity and strengthen competitiveness to gain ground in the global market. An increased market share can then generate faster growth and create more jobs. Therefore, catch-up growth sustained by technological progress and productivity growth is the fundamental solution to unemployment and poverty alleviation.

The Competitiveness Challenge

A recent Industrial Development Report of the United Nations Industrial Development Organization (UNIDO 2009) provides a clear picture of the global competition facing Sub-Saharan Africa's manufacturing industries. As suggested by the data in table 1.1, African manufacturing industries are squeezed by competition pressure on two fronts. On one front, Africa faces competition from knowledge-intensive economies in the industrialized countries, which occupied 70 percent of the global manufacturing market in 2005. On another front, Africa must compete with other developing and transition economies, including the emerging manufacturing giants in Asia, which derive much of their competitive strength from low labor costs. Together, other developing and transition economies enjoyed a market share of over 20 percent in 2005, leaving only 1 percent for Sub-Saharan African firms, of which South Africa alone accounts for more than half.

Data of South African industries[2] suggest that southern African manufacturing industries may have very limited competitive advantage in terms of labor cost in their competition with Asia. As illustrated in figure 1.2 (Mengistae 2011), South Africa can derive little competitiveness from

Table 1.1   Composition of Global Manufacturing Value Added, 2005

Economies	Share (%)
Industrialized countries	69.4
South Africa	0.4
Sub-Saharan Africa excluding South Africa	0.3
China	9.8
East Asia and the Pacific excluding China	7.7
Mexico	1.7
Latin America and the Caribbean excluding Mexico	4.7
Middle East and North Africa	2.2
Transition economies	1.7
India	1.4
South Asia excluding India	0.4
Least-developed countries	0.3

Source: UNIDO 2009.

Figure 1.2 Unit Labor Costs in Selected Industries, 1996–2007

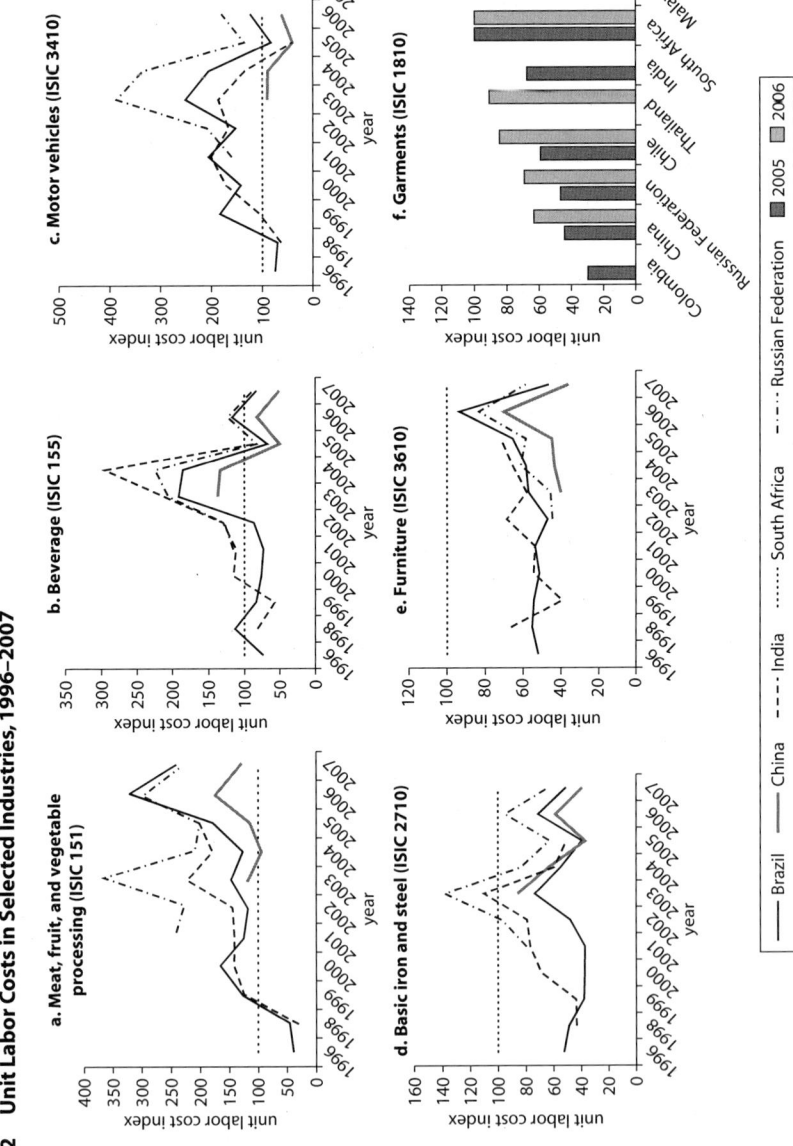

Source: Mengistae 2011.
Note: Index, South Africa = 100. ISIC = International Standard Industrial Classification.

unit labor cost in sectors such as furniture, garment, steel, and beverage manufacturing. In the motor vehicle sector, South Africa used to enjoy an advantage of lower unit labor cost over countries other than China. This position seems to have changed, however, because these countries have been catching up since 2003. Of the six sectors studied, meat, fruit, and vegetable processing appears to be the only sector where South Africa has a clear advantage in lower cost.

To the extent that South African manufacturing industries are representative of manufacturing industries in southern Africa, the implication is that southern African economies must rely on increasing productivity to effectively compete with both knowledge-intensive and low-labor-cost economies. Of course, one cannot say that competitiveness derived from low cost is not important to southern Africa. Indeed, southern African industries have significant potential to strengthen competitiveness through better management of costs of labor, infrastructure, and other inputs. However, in comparison with low-cost Asian emerging economies, southern Africa needs to gain more in terms of overall efficiency to compensate for its high labor and other costs. In other words, technological progress and total factor productivity are more important to Sub-Saharan Africa's manufacturing industries than they were to emerging Asian economies.

Productivity

Efficiency and productivity do not appear to be the strengths of southern African economies, however. Although the lack of available data does not allow a comprehensive assessment, analysis of the trend of South African manufacturing industries indicates clearly that labor productivity growth, albeit strong, is driven primarily by rising capital intensity (figure 1.3).

South Africa has globally leading technologies in mining and mining-related activities, and it has a significant number of high-quality patents. Firms in mining equipment develop much of their intellectual property locally. South Africa is a world leader in a number of mining equipment products, including spirals for washing coal and equipment for pumping up water, hydropower, tracked mining, underground locomotives, ventilation, shaft sinking, and turnkey new mine design, among others. South African expertise is particularly advanced and at the global frontier in deep-level mining and associated competencies. However, as can be seen in chapter 2, skill shortages have been threatening South Africa's global technological leadership in this field. Firms reported skill shortages at all levels from managerial and engineering positions to welders and

Figure 1.3   Trends in Labor Productivity and Capital: Labor Ratio in the South African Manufacturing Sector, 1970–2008

Source: Quantec database.
Note: Index, 2005 = 100.

boilermakers. And this undersupply is being exacerbated by aggressive recruiting on the part of competitor firms, especially in Australia, which has been very successful at recruiting skilled South Africans. In addition, given the relative success of mining-related capital equipment exports linked to South African engineering and project management capabilities, whether the spillover potential from this sector to the rest of the economy has been fully explored is not clear.

Exports

The weakness of southern African manufacturing industries in terms of productivity and competitiveness is also reflected in their export performance. Lack of progress in economic diversification has resulted in continuous dependence of southern Africa's export basket on a small number of primary products. A close look at the export performance of the four southern African countries points to the limited technology content of their exports.[3] In all four countries, exports are dominated by a combination of primary commodities and low-technology manufactures. The detailed analysis is provided in the annex to this chapter. The analysis shows that exports from South Africa are dominated by medium-technology goods, which accounted for 19 percent of total merchandise exports in 2009. This structure has been steady since 2001, suggesting that in this period South Africa's export basket has shown little diversification. Overall, the exports of low-technology, medium-technology, and high-technology goods account for approximately 30 percent of South Africa's total merchandise exports.

Mauritius's exports are predominantly garments and apparel products.[4] Low-technology goods account for 39 percent of total merchandise exports in 2008. Sugar and sugar confectionary constitute the second-largest segment of exports, accounting for approximately 11 percent of total exports. The remainder is fragmented across a broad range of products, including processed food and electrical machinery and equipment.

Namibia's exports comprise mainly agricultural and resource-based products. As a result, technology-intensive exports accounted for only 9 percent of total merchandise exports in 2008. Lesotho's exports are heavily concentrated in textiles, which fall into the low-technology category, despite some resource-based products. However, the share of low-technology exports in total merchandise exports declined from 98 percent in 2004 to 56 percent in 2008. This drop reflects the end of the Multifiber Agreement (MFA) on Textiles and Clothing and increased competition from Asian exporters of clothing and textiles in lucrative markets, particularly the United States, and an increase in resource-based manufactures driven mainly by precious stones.

FDI

The sectoral split of FDI into a country may also affect technology and growth. Wang (2009) suggests that most of the recorded positive impact from FDI comes from investment in the manufacturing sector and that nonmanufacturing FDI (agriculture, services, and mining) plays an insignificant role in promoting growth. A possible reason may be that the manufacturing sector includes both labor-intensive industries (food processing and textiles) and research and development (R&D)-intensive industries (computers and electrical machinery), whereas other sectors do not. This factor has implications for Sub-Saharan Africa, especially for those countries whose competitive advantage tends to be in nonmanufacturing sectors.

South Africa's top two FDI destination sectors in 2007 were mining and business services. Total inflows into these sectors amounted to US$13,944 million, dwarfing the manufacturing figure of US$2,410 million. A similar scenario is evident in Mauritius, where the key sectors are tourism, banking, and real estate. Inflow into these sectors totaled US$346 million in 2007, compared with manufacturing's US$9 million, limiting the extent of potential technology transfer.

A comparative overview of FDI flows and stocks into the four countries is presented in table 1A.2 in the annex. Table 1A.3 contrasts the stocks and flow figures with those of a few comparator countries.[5] Except for Lesotho, the majority of FDI comes from Europe and the United States. However, most of the FDI is concentrated in mining (South Africa) and services (Mauritius). Although flows for most of the countries rose during 2006–09, they are still relatively small as a percentage of gross domestic product (GDP), especially in the manufacturing sector. These trends in the four countries under review suggest that exports are to a large extent resource driven, and FDI has been flowing into services rather than manufacturing.

The Relevance of Technology Absorption

To sum up, the overall pattern of global competition facing southern African manufacturing industries has a clear implication: the future of southern African manufacturing industries depends critically on technological progress and productivity growth. The less advantage an economy has in terms of low labor cost, the stronger is such dependence. As argued earlier, in southern Africa, as in all other developing economies, technology transfer and absorption is the primary route toward technological progress, despite the importance of technology innovation. The rationale of this book lies here.

But is technology absorption or, more generally, technology progress, relevant to job creation, which is the top priority of South Africa as well as other southern African countries? Although technology progress is often associated with reduced demand for low-skilled labor (see Sanders and ter Weel [2000], for some of the empirical evidence of skill-biased technical change), concluding that technology progress is irrelevant to job creation would be wrong. First, for many southern African firms, technology progress is a necessity to remain globally competitive, simply because their competitors are improving technology and productivity. Such progress is therefore essential to maintain existing jobs, before getting to the agenda of job creation. More generally, technology progress is essential to productivity, which is a key determinant of competitiveness and hence the share of an economy in the global market. It is therefore also essential to growth. Although employment elasticity of growth (that is, the percentage change of employment that results from every 1 percent of GDP growth) can vary on the growth path of a country, studies comparing South Africa's postapartheid economic growth and employment data with those of Brazil, the Russian Federation, India, and China (BRIC countries) suggest that faster growth is the first-order challenge to South Africa. Indeed, despite the persistence of high unemployment, postapartheid economic growth in South Africa generated a rise of total employment by 26 percent from 1995 to 2008. Every 1 percent GDP growth during this period was associated with 0.43 percent growth in employment. This figure represents an employment elasticity of GDP growth that is significantly higher than that of Brazil (0.30 percent), Russia (0.10 percent), India (0.20 percent), and China (0.06) in the same period.[6] The job creation potential of GDP growth in South Africa is large. What is required is faster growth, which can be realized only by enhancing competitiveness in the global market.

The Determinants of Technology Absorption: A Conceptual Framework

The definition of technology transfer that was agreed on during the negotiations of the International Code of Conduct for Transfer of Technology is "the transfer of systematic knowledge for the manufacture of a product, for the application of a process or for the rendering of a service and does not extend to the transactions involving the mere sale or mere lease of goods" (UNECA 2010, 1). *Technology absorption* refers to activities undertaken by the receiving entities of technology transfer. It can be defined as a learning activity that a firm can employ to integrate and commercialize technological knowledge that is new to the firm and transferred to it, or acquired by it, from outside (Goldberg and others 2008). Three key factors determine the outcome of technology absorption activities. First of all, technology absorption will not happen if the firm has no incentive to invest in it. If one assumes the existence of incentives, key factors that determine the outcome of technology absorption activities are access and capacity (World Bank 2008).

Incentives

Before technology absorption can happen, firms at the receiving end must have incentives to invest in technology transfer and absorption. This requirement is likely a function of many broader factors: in particular, intense and fair product market competition, both domestic and foreign, that would push firms toward greater efficiency and business productivity. Other important factors include policies and regulations that reduce the costs and risks of doing business and promote access to finance. Rodrik (2008) suggests that innovation requires rents, without which entrepreneurs do not have the incentives to invest in discovery. This factor applies to technology absorption wherever similar profiles of risks and payoffs are present.

Market structure and competition are likely to affect a firm's incentive to innovate or absorb technology. To the extent that the knowledge developed through investment in R&D spills over to, or is absorbed by, other firms, absorption may create disincentives to innovation in certain market structures. In theory, the relationship between competition and innovation by incumbent firms is ambiguous (Aghion and others 2005). The threat of a firm with new or higher-quality products can affect incumbents' incentives to invest in more R&D and other investments to introduce new products (Lederman 2010). Firms that are far behind the technological frontier may choose to reduce investments in innovation in the face of competition from new entrants, because innovation is very costly to them and competition would erode rents obtained from innovating. By contrast, firms that are close to the technological frontier need to spend relatively little to stay ahead of new entrants, and therefore competition would create greater incentives to spend on innovation. Openness to trade in general is likely to increase exposure of domestic firms to foreign competition and induce the adoption of more advanced technologies in both export- and import-competing firms (see Schiff and Wang 2006).

Aghion, Braun, and Fedderke (2008) demonstrate that in the case of South Africa, a positive effect of competition on productivity growth occurs in the manufacturing sector; therefore, the "escape competition" effect on innovation described above seems to dominate disincentives to innovate that are linked to competition in that country. These authors argue that markups are significantly higher in South African industries than they are in corresponding industries elsewhere in the world and that a 10 percent reduction in markups in South Africa would lead to an increase in productivity growth of 2 percent to 2.5 percent per year.

The incentives for firms to absorb technology will also depend on the extent to which the *overall investment climate* in a country is favorable. In particular, much depends on whether firms investing in technology transfer and absorption can appropriate the benefits of those investments, whether the infrastructure is of good enough quality to allow for the profitable undertaking of the activities, and whether the financial system is sufficiently deep to provide adequate resources to finance the costly process of technology absorption. Furthermore, the quality of the investment climate might affect the technology transfer between multinational firms and their subsidiaries.

Access

Well-motivated firms must have access to technology before they can absorb it, and access often is seen as a mirror image of barriers. Barriers or boundaries with regard to technology transfer typically exist between firms and countries. Barriers between firms often are established because of the ownership of the technology in question by one firm and the lack of it by another. The boundary between countries is an additional barrier when technology is transferred from a firm in one country to another firm in a different country. However, technology transfer can obviously be intracountry, intraindustry, or even intrafirm. Getting access to technological knowledge, therefore, often involves various contractual arrangements, which may not always be affordable to the firms at the receiving end. Incentives for technology owners can be instrumental in increasing access to the technological knowledge they possess, particularly tacit knowledge, that is, technological knowledge that cannot be codified and must be absorbed through "learning by doing."

As for other kinds of knowledge, the knowledge of technology, when it is transferred, can be carried by paper and electronic media such as written text (for example, a manual), graph (a blueprint), and electronic file (software). It can also be carried by the human brain, machinery and equipment, and goods. The implication is that all movements of technological knowledge carriers, including, for example, the movements of personnel, services they provide, machinery and equipment, and other goods, may lead to technology transfer. This assumption gives rise to three key channels of technology transfer (Hoekman and Javorcik 2006).

The first channel is trade in goods and services. All trade bears some potential for transmitting technological information. Imported capital goods and technological inputs can directly improve productivity by being used in production processes. Alternatively, firms may learn about technologies by exporting to knowledgeable buyers, who share product designs and production techniques with them. Lederman (2009), using firm-level data from enterprise surveys in several countries, finds that a firm's export status (that is, whether a firm exports more than 10 percent of its sales) is positively correlated with the probability that a firm innovates. Historical accounts of the rise of East Asian export industries stress the role played by advanced-country buyers as conduits of technological and managerial know-how to developing-country firms (Pack and Westphal 1986). For example, China engaged in aggressive technology importation for two decades before shifting its focus to indigenous innovation in the mid-2000s when it was established as the world's workshop (Zhang and others 2009).

Similarly, case studies and interviews suggest that the cultivation of manufacturing outsourcing relationships between local firms and multinationals—with or without a local presence in the market—can also serve as a channel of knowledge flows (Pack and Saggi 2001). Empirical studies suggest that international trade mediates flows of knowledge, allowing firms and industries to acquire technologies that expand their productive capabilities, often in ways that show up in conventional productivity measures (see Coe and Helpman 1995; Keller 2004).

Similar to trade, FDI has the potential to be a crucial channel for technology transfer and absorption. FDI can lead to technology absorption by transferring knowledge about new technologies from foreign firms to domestic subsidiaries or simply through the know-how and equipment brought in by foreign investors. One of the reasons FDI is considered an important mechanism for technological development is that multinational firms possess much of the world's stock of technological knowledge. This mechanism seems to have operated in Mauritius, where several firms located in export processing zones were owned by employees who had received on-the-job training in foreign enterprises (World Bank 2009).

A number of potential channels for FDI spillovers are generally identified (Blomström and Kokko 2003; Saggi 2002; UNCTAD 1995): *demonstration effects*, in which technology is transferred through imitation or reverse engineering; *labor turnover*, in which technology is transferred through workers; and *vertical links*, in which technology is transferred when multinationals interact with their immediate suppliers and buyers. Kinoshita (1998) adds a *competition effect* and a *training effect* to the preceding. The competition effect refers to the entrance of foreign firms causing a reduction in monopoly profits in local firms and forcing them to adopt the latest technologies to retain market share. The training effect refers to local firms training their own workers to increase product quality and remain competitive. Kinoshita (1998) notes that this local accumulation of skills is often necessary to absorb the new technology. Furthermore, because the skill is often specific to the technology, the cost of the training needs to be taken into account. She also highlights the difficulties in separating the various effects, because they interact to a substantial degree. Empirical studies tend to validate the presence of knowledge spillovers, and most also find a significant growth effect (see Coe and Helpman 1995; Eaton and Kortum 1996; Xu and Wang 1999).

The role of FDI is further highlighted by the *global engagement hypothesis*, which suggests that firms engaged in global business activities through foreign investment, adoption of foreign technologies, or imports of capital goods are more likely to undertake innovations than other firms (Lederman 2009). This hypothesis on the role of FDI is further corroborated by the *information spillovers hypothesis*, which suggests that firms that have broader access to commercial knowledge, including by adopting foreign technologies, will tend to have higher propensities to innovate than otherwise similar firms (Lederman 2009).

The third channel of technology transfer is direct trade in knowledge through technology purchases or licensing. It may occur within firms, among joint ventures, or between unrelated firms. Licensing and FDI are often substitutes, but they may also be complements. Much of the recorded international payments and receipts for intellectual property (royalties) occur within firms as flows between parent firms and affiliates. Which form is preferable to technology owners depends on many factors, including the strength of intellectual property rights (Hoekman and Javorcik 2006).

The choice of the channel through which a technology is transferred is a function of many factors. The nature of the technology is a key determinant. For example, some technologies are more codified than others or easier to copy than others. The business strategy of the technology owner and buyer is often a major determinant. For instance, drug producers often keep the production and process technologies within their networks of firms while licensing mainly the products (patents of drugs, vaccines, and equipment). Although every drug producer knows how to make aspirin, few can produce it at a competitive market price. The difference may be in the production and process technologies.[7] The capacity of the technology buyer and the technology seller may also influence the modes within which transfer of technology may occur. For instance, if the buyer has excellent knowledge of the technology and needs only access to the technology, licensing may be preferred. This factor is also important to the technology seller, as the ability of the buyer to successfully bring the technology to market, achieve economies of scale, and compete favorably in the national or global marketplace is necessary because license fees are often based on the proportion of sales and profits generated by the buyer (UNECA 2010).

Although the discussion of technology transfer and absorption in this book concentrates mostly on transfers from the global technology pool to firms in southern African countries, technology transfer and absorption also take place within one country, one industry, or even one firm. Indeed, intraindustry technology diffusion has been found to be critical to raising industrywide productivity in some manufacturing sectors in India and China (Dutz 2007; Zhang and others 2009). Known for their positive externality, R&D activities by nature spread knowledge among related firms and individuals. What is viewed as "leakage" from one perspective can be "diffusion" from another.

Capacity

The absorption of existing technology through trade, FDI, or licensing is neither automatic nor without cost (Cohen and Levinthal 1989; Kinoshita 1998). Motivated firms that are exposed to more advanced technology would not be able to draw any benefits if they are constrained by their own absorptive capacity. For a country to be able to reap the benefits of increased access to international technological knowledge, it requires some minimum prerequisites. The ability of firms to absorb technology depends to a large extent on the skill levels of its managers and its labor force. In addition, R&D has been proved to be increasingly important to technology absorption in addition to "new to the world" innovations. Industry research links would be key here, as would links to the developed countries, with an increasing trend toward coinventions and collaborations.

Skills of the workforce or human capital in general are of fundamental importance to technology absorption for a simple reason: all activities of technology absorption must be conducted by people, and all knowledge can be "absorbed" only by people. An educated workforce can therefore be considered a precondition for a country to have the capacity for knowledge acquisition and adaptation, especially in an environment where firms face competitive pressures that call for frequent changes in the product mix and the production technology. Skilled workers with engineering and scientific competencies are required to make significant adaptations of existing technologies or to create new ones (De Ferranti and others 2003). For example, firms cannot perform reverse engineering of high-tech products created abroad if they lack an adequate supply of engineers in the first place or if a skill premium makes their use in the production process too costly. Lederman (2009) presents empirical evidence confirming that the probability that a firm innovates is linked to the level of education of its personnel.

The importance attached to absorptive capacity as a factor underlying growth is also stressed by Leipziger in Chandra (2006, xiii), arguing that governments in developing countries can make a significant contribution to growth by "encouraging or enabling the private sector to seek, acquire, adapt, and deploy modern technologies of production while investing in the skilled workforce that is vital for technological adaptation and building the dedicated institutions that will foster the process of modernization."

Education hence provides the foundation for innovation and technology absorption. As the *Global Competitiveness Report 2010–11* reiterates: "Quality higher education and training is crucial for economies that want to move up the value chain beyond simple production processes and products. In particular, today's globalizing economy requires countries to nurture pools of well-educated workers who are able to adapt rapidly to their changing environment and the evolving needs of the production system" (WEF 2010, 5).

Examples include the three learning economies—Finland, Ireland, and Singapore—that are studied by Yusuf and Nabeshima (2011). Analyzing the factors the economies had in common to explain their sustainable high growth rates, the authors found that all three had stressed quality of education at all levels of schooling, giving specific importance to science, technology, engineering, and mathematics; increased access to higher education; and investment in research universities and research capacities of the economy at large.

What, then, is the route to successful skill development? A recent World Bank (2010) study on stepping up skills proposes a conceptual framework known as STEP (Skills Toward Employment and Productivity), which focuses on five interlinked steps:

- *Getting children off to the right start*—by developing the technical, cognitive, and behavioral skills conducive to high productivity and flexibility in the work environment through early child development and by emphasizing nutrition, stimulation, and basic cognitive skills.
- *Ensuring that all students learn*—by building stronger systems with clear learning standards, good teachers, adequate resources, and a proper regulatory environment.
- *Building job-relevant skills that employers demand*—by developing the right incentive framework for both preemployment and on-the-job training programs and institutions (including higher education).
- *Encouraging entrepreneurship and innovation*—by creating an environment that encourages investments in knowledge and creativity.

- *Matching the supply of skills with the demand*—by moving toward more flexible, efficient, and secure labor markets.

All stages of skill development involve service delivery. From this perspective, a number of cross-cutting factors can be identified that are crucial to the results (World Bank 2004, chap. 3). First, all service providers must have the right incentives to respond to demand and to deliver services efficiently. This factor often takes a governance structure that provides a proper balance between autonomy and accountability. Second, service providers must have access to adequate financial resources to fund the operations they need to deliver their services. Third, they must have the capability to deliver, in particular, to be supported by an adequate supply of human resources. Finally, the government must play a proper role to mitigate market failure and ensure that services are delivered equitably. Government financing is a common instrument to ensure equitable access to services and overcome inefficiencies arising from public goods and externalities. Direct provision of services by public institutions and government regulations are also frequently used by the government to ensure satisfactory delivery of services.

A framework such as the one in figure 1.4 can be used to guide diagnostic analysis.

Although skill development is often discussed in the context of the national economy, it is increasingly internationalized, especially with regard to entrepreneurial or business skills and R&D or engineering skills, which are more mobile. The global competition for talent, which normally results in "brain drain" in poor countries and "brain gain" in rich countries, reinforces the technological gap between them. However, many developing economies have managed to turn the tide. For example, the high-skilled diaspora of countries such as India has contributed to the growth of the information technology (IT) and outsourcing sectors. Overseas Taiwanese engineers and returnees have worked closely with policy makers to establish a successful venture capital industry, helping finance high-risk entrepreneurial activities in the technology sector. The impact of global competition on talents is particularly significant for small countries where economies of scale would create extra cost if countries try to educate talents in every discipline in domestic universities.

Domestic R&D capacity is a crucial determinant of the ability to effectively absorb technology. The collaboration of R&D efforts among firms in developing countries and industrialized economies is becoming more

Figure 1.4   Skill Development: An Analytical Framework

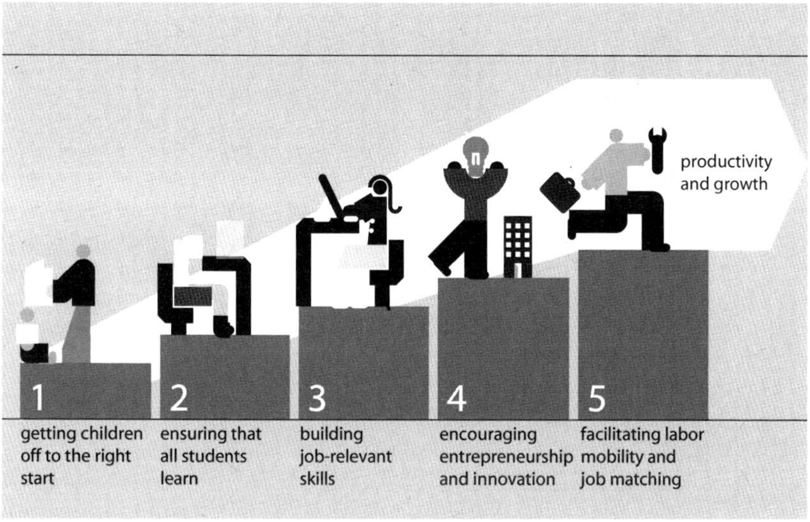

productivity and growth

1	2	3	4	5
getting children off to the right start	ensuring that all students learn	building job-relevant skills	encouraging entrepreneurship and innovation	facilitating labor mobility and job matching

Source: World Bank 2010.

frequent, serving as a crucial channel of absorption.[8] Findings of a survey of industries across 12 countries of the Organisation for Economic Co-operation and Development by Griffith, Redding, and Van Reenen (2004) show that R&D enhances technology transfer by improving the ability of firms to learn about advances on the technology frontier. Thus, R&D is important both in the process of catching up and in directly stimulating innovation. Cohen and Levinthal (1989) refer to this as the "second face" of R&D.

In countries in southern Africa, R&D investment by the business sector is usually low, with weak links between industry and academic research. Academic curricula frequently do not respond to the needs of the industry. South Africa is an exception, with significant resources devoted to R&D (over 50 percent of which are funded by businesses) and a small number of large firms that are close to the world technological frontier but that do not present strong links to the rest of the economy. Nevertheless, even in South Africa, an inherent disconnect persists between industry and academic research (see Kaplan 2008).

Collaboration between industry and research as a result of the free flow of knowledge is also vital for effective technology absorption capacity. Systemic innovation theory has given rise to the insight that technological innovation is the result of a division of innovative and creative labor (Freeman 1988; Lundvall 1992; Nelson 1993). Interdependent exchanges between different organizations, such as private firms, universities, research laboratories, suppliers, and customers, can result in effective technological innovations. The rationale for this collaboration rests on the lack of sufficient incentives for individual firms to undertake uncertain and imperfectly appropriable research at the socially optimal level. Some of the market failures impeding investments in R&D could be considered targets for economic policy prescriptions, if the possible market failures more than outweigh the expected government failures (Racine and others 2009).

The Role of the Government

The rationale for government intervention in innovation is based on two main sources of market failures: (a) partial appropriability, which does not allow inventors to capture all the benefits of their invention; and (b) information asymmetries, for example, the difference between the information that an inventor looking for financing has about an invention and the information that the potential financier has (Goldberg and others 2006). But does a rationale also exist for intervention to support technology absorption rather than innovation?

Several arguments can be made in favor of government support for technology absorption. Hausmann and Rodrik (2003) reason that countries lagging behind the technology frontier will underinvest in technology absorption because of the shortcomings of their intellectual property regime. Unlike innovating companies, which are protected by patents, entrepreneurs who invest funds to discover which technology to adapt to a developing-country context do not normally get any protection for the markets they open, no matter how high the associated social return may be. The right way of thinking of absorption, Rodrik (2004) suggests, is as a discovery process in which both firms and the government engage in strategic coordination while learning about underlying costs and opportunities. Here, Rodrik argues, the fundamental departure of this viewpoint from classical trade theory is that entrepreneurs may lack information about where the comparative advantage of a country lies or, at the micro level, entrepreneurs may simply not know what is profitable and what is not.

Lin and Monga (2010, 24) highlight this coordination between the government and firms: "High-performing developed and developing countries are those where governments were able to play an active role in the industrial upgrading and diversification processes by helping firms take advantage of market opportunities. They have generally done so by overcoming the information, coordination, and externality issues and by providing adequate hard and soft infrastructure to private agents." Through a set of extensive case studies of Bangladesh, India, Tanzania, and Thailand, Khan (2009) shows that the emergence of competitive success in their most important high-growth sectors depended not just on access to the market in a liberal economy but much more fundamentally on the factors that combined specific financing instruments with appropriate mixes of governance, market, and political factors that ensured high levels of effort and increased competitiveness. He argues that without specific policies to develop technological capabilities, the market by itself will not pull sectors, regions, and countries with low capabilities toward sustained development.

The Empirical Methodology of This Book

The empirical analysis in this book, which is detailed in chapter 2, comprises case studies of various sectors in four countries, Enterprise Survey (ES) data for the four countries, and National Innovation Survey (NIS) data for South Africa. The book uses the findings of econometric analyses to complement some of the findings that were arrived at from the case studies to shed light on various channels of technology absorption. The econometric analyses use two data sources: the South Africa NIS of 2005, conducted by the Human Sciences Research Council (HSRC) for the Department of Science and Technology, and the ESs for Lesotho, Mauritius, Namibia, and South Africa, conducted by the World Bank.

Profile of the Case Studies Analyzed

The case study analyses aimed to determine the extent of technology absorption taking place in exporting firms. Firms that engaged significantly in exports were selected because the demands to enhance technology are particularly strong for exporters. The study analyzed the motivation for those exporters to undertake activities that increase technology absorption and the constraints faced by the exporters to increase the scale of the technology-upgrading investments. To gauge the level of technology absorption taking place among foreign-owned firms, the sample in Mauritius consisted of a few foreign-owned exporting firms, and the sample in Lesotho consisted of all foreign-owned exporters. Detailed, semistructured interviews were undertaken with top management. The firms were visited and, in most cases, the visit included a tour of the factory. In addition, where possible, interviews were conducted with exporters associations.

South Africa. Three sectors classified as medium technology (auto components, capital goods, and chemicals) and one high-technology sector (software and IT) were selected for an in-depth analysis. Exporting firms from each of these sectors (including large, medium, and small firms) were interviewed to gain insights into how technological adaptation enables exports and how, in turn, exports contribute to the enhancement of technological capacities.

Namibia. The agroprocessing sector in Namibia makes up more than half of Namibia's exports; hence all the exporters interviewed in Namibia were in the agroprocessing sector. Four firms were interviewed: two large fisheries, a meat processor, and a brewery. The brewery and the meat processor are monopolies and therefore represent the entirety of their respective industries. The two fish processing firms are two of the dominant firms in the fish processing industry and together are responsible for approximately a quarter of employment in the fisheries sector.

Mauritius. Sixteen exporting firms that had been surveyed in the ES for 2008 were selected to gain insight into what they considered as their new product or new process innovation and the constraints they face for greater investments in technology absorption. Of the 16 firms, 5 were foreign owned. Fourteen firms were in the manufacturing sector that included textile and garment firms, chemicals, food and beverages, and capital equipment, and 2 were IT firms.

Lesotho. Ten textile and garment firms, 5 each from the Maseru and Maputsoe regions, were interviewed. All 10 firms, in addition to being exporting firms, are foreign owned. Firms in Maseru are Taiwanese owned; firms in Maputsoe are South African owned.

The South Africa NIS

The 2005 NIS in South Africa represents the first attempt to develop a survey instrument based on a framework similar to the European Community Innovation Surveys.[9] The hope is that this instrument will be developed over time and administered more regularly in the future to build up the country's microdata sources for firms.

The NIS was designed and implemented by the HSRC of South Africa in conjunction with Statistics South Africa. The survey was based on a stratified sample (by sector and size of enterprise) drawn from a business registry database housed in Statistics South Africa. The number of firms contacted was 3,087, but the final realized sample was only 979 firms, representing a response rate of approximately 37 percent (Blankley and Moses 2009, 6).[10] Unfortunately, the surveys ask only a few questions of firms that introduced neither a new product nor a new process. All that is known about these "noninnovator" firms is their turnover, exports, and number of employees (in levels and growth rates); the main industry they belong to; and their potential affiliation to a group. The 979 firms surveyed can be classified into four categories: (a) companies that introduced a new product—515 firms; (b) companies that introduced a new process—486 firms; (c) companies that introduced both a new product and a new process—396 firms; and (d) companies that introduced either a new product or a new process—605 firms ("innovators").

The NIS data set contains a number of variables describing outcomes closely related to technology absorption, each of which is obtained from detailed interviews of managers of firms. In these surveys, firm managers are asked specifically whether their firm recently introduced a new product (either new to the firm or new to the market), acquired a new production process, or other. Each of these potentially represents a dimension of the technology absorption process.

The NIS also allows the comparison between new-to-the-market and new-to-the-firm innovations. The analysis characterizes the relationship between a firm's probabilities of introducing any of the dimensions of technology absorption and the underlying determinants of absorption outcomes by estimating a production function for technology absorption:

Technology absorption = f ($K, L, OPEN, IP, R\&D, HC, IC, PS$),

where K is access to finance, L represents firm employment, $OPEN$ represents the openness of the firm to international trade and knowledge from abroad, IP is the ownership of intellectual property (patents and know-how), $R\&D$ indicates whether the firm conducts R&D, HC refers to human capital, IC is a catchall variable for the investment climate, and PS is public support for technology absorption. In this formulation, the determinants can be seen as inputs into this production function. All regressions control for the firm's sector and size, as measured by total employment.

To identify the determinants of technology absorption, correlations are estimated (using multivariate regressions) between the measures of technology absorption and a variety of explanatory variables, which include firm employment, R&D, firm export status, foreign ownership or FDI status, tertiary education, subsidiary status, financial constraints, an aggregate intellectual property rights indicator or patents in South Africa, patents applied for outside of South Africa, registration of an industrial design, and registration of a trademark. The term *correlation* is used in reference to the relationship of technology absorption dimensions with the determinants or inputs into the production function because the direction of causality between the dimensions and the determinants cannot be identified. A positive correlation of technology with export status, for instance, cannot suggest whether the technological capacity causes a firm to become an exporter or, alternatively, exporters increase their technological capacity because of market requirements.

Description of the ES Data

The ESs for the four countries vary in terms of sample size. South Africa has the largest sample, with 1,056 firms, and Lesotho has a small sample size of 75 firms. Table 1.2 shows the share of firms that export their output either directly or indirectly through suppliers and the share of firms that have foreign ownership. Namibia has the largest share of exporting firms (17 percent), and Lesotho has the largest share of firms with foreign ownership. Detailed country analysis was conducted for only South Africa and Mauritius because of the small sample sizes for Lesotho and Namibia.

The surveys examine a wide range of factors affecting the business environment. These data possess many indicators of the factors that would enhance the underlying absorptive capability of firms, including, among others, the skill mix of the workforce, training levels, R&D spending, and measures of international connectedness. The data sets contain a number of variables describing outcomes of technology absorption. These include the introduction of a product that is new to the firm, the upgrading of an existing product, the acquisition of a new production technology, the acquisition of new product licensing agreements, the acquisition of a new quality certification, variables related to

Table 1.2   Firm Characteristics

Country	Sample size	Exporter firm (%)	Foreign ownership (%)
South Africa	1,056	11.28	12.88
Namibia	106	16.98	26.42
Mauritius	384	14.74	10.67
Lesotho	75	60.00	49.33

Source: Enterprise Surveys (database), World Bank, Washington, DC. https://www.enterprisesurveys.org.
Note: Exporter firms are defined as firms that export more than 10 percent of sales. *Foreign ownership* is defined as firms that have more than 10 percent private foreign ownership.

"soft innovation," and technology use. One can view these variables as multiple indicators of an underlying latent variable: the absorptive capacity of the firm. Although this capacity cannot be directly measured, factors that contribute to it and outcome variables that are related to it can be measured. These outcome variables are the dependent variables used to investigate the relationship between absorptive capability of firms and actual absorption outcomes.

For South Africa, the 2003 ES covered a sample of 800 firms, of which only 587 are in the manufacturing sector. The sample was collected across four regions in South Africa: Cape Town, Durban, Johannesburg, and Port Elizabeth. In term of sectors, the sample of firms covered seven economic sectors: beverages and food; garments, leather, and textile; furniture and wood; paper and printing; chemical products; equipment and machinery; and motor vehicles. The 2007 ES covered a sample of 1,056 firms. It sampled firms in Cape Town, Durban, Johannesburg, and Port Elizabeth operating in the following economic sectors: manufacturing, construction, retail and wholesale service, hotels and restaurants, transport, storage and communication, and computer-related activities. In the 2007 ES, not all firms were asked the questions on innovation. Therefore, only those firms that responded to either the product or process innovation question are included in the sample for this book, resulting in a final sample size of 427. The 2008 ES for Mauritius has a total sample of 384 firms, with 177 manufacturing firms and 207 firms in the service industry. As in the NIS, the analysis conducted characterizes the relationship between a firm's probability of introducing a new or significantly improved product or process and the underlying determinants of absorption outcomes, by estimating a technology absorption production function.

Annex

The technological content of exports and imports in the four countries was reviewed using the technology-based product classification developed by Lall (2000). Lall's approach is based on the second revision of the Standard International Trade Classification of commodities in which products are classified into primary products, resource-based manufactures, low-technology manufactures, medium-technology manufactures, and high-technology manufactures. Consequently, three technology categories (low, medium, and high) are expressed as percentages of total merchandise (not just manufactured) exports, in conformity with Lall's approach (see table 1A.1). The detailed analysis is provided in this annex.

Table 1A.1   Technological Classification of Merchandise Trade

Classifications	Examples
Primary products	Fresh fruit, meat, rice, cocoa, tea, coffee, wood, coal, crude petroleum, gas
Manufactured products	
Resource-based manufactures	
Agro- and forest-based products	Prepared meats and fruits, vegetable oils, beverages, wood products
Other resource-based products	Ore concentrates, petroleum and rubber products, cement, cut gems, glass
Low-technology manufactures	
Fashion cluster	Textile fabric, clothing, head gear, footwear, leather manufactures, travel goods
Other low technology	Pottery/simple metal parts/structures, furniture, jewelry, toys, plastic products
Medium-technology manufactures	
Automotive products	Passenger vehicles and parts, commercial vehicles, motorcycles and parts
Process industries	Synthetic fibers, chemicals and paints, fertilizers, plastics, iron, pipes/tubes
Engineering industries	Engines, motors, industrial machinery, pumps, switchgear, ships, watches
High-technology manufactures	
Electronics and electrical products	Office/data processing, telecommunication equipment, TVs, transistors, turbines, power generating equipment
Other high technology	Pharmaceuticals, aerospace, optical/measuring instruments, cameras
Special transactions	Electricity, cinema film, printed matter, "special" transactions gold, art, coins, pets

Source: Lall 2000.

Structure of South Africa's Trade in Technology-Intensive Goods

Technology exports. Figure 1A.1 gives the relative structure of South Africa's exports shown by technology content from 2001 to 2009. Exports were dominated by *medium-technology goods*, which accounted for 19 percent of total merchandise exports in 2009. In 2009 exports of low-technology and high-technology goods accounted for 3 percent and 9 percent of total merchandise exports, respectively. This structure has been steady since 2001, suggesting that in this period South Africa's export basket has had little diversification. Overall, the exports of low-, medium-, and high-technology goods together account for approximately 30 percent of South Africa's total merchandise exports.

Figure 1A.1   South Africa's Exports of Low-, Medium-, and High-Technology Goods, 2001–09

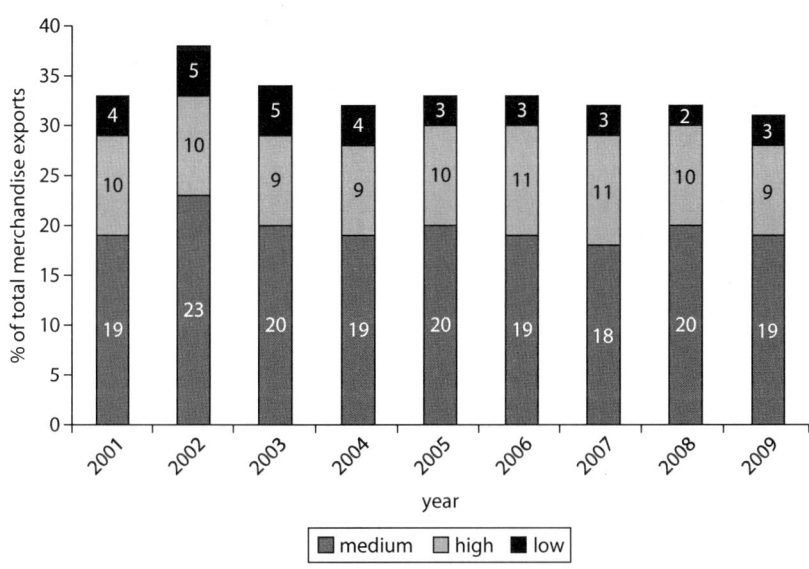

Source: South African Institute of International Affairs (SAIIA) calculations based on International Trade Commission (ITC) data, 2010.

Figure 1A.2 shows South Africa's value of technology exports from 2001 to 2009. The value of total exports since 2001 reached a peak in 2008. The decline after 2008 is a result of the global recession. In the period before the crisis, medium- and high-technology exports experienced the highest growth, while low-technology exports remained relatively stable. The growth in medium-technology manufactures reflects an increase in motor vehicle–related exports, largely as a result of the Motor Industry Development Program introduced in 1995.

Technology imports. The structure of South Africa's merchandise technology imports is similar to that of its merchandise exports, which is dominated by medium-technology and high-technology manufactures. In 2009, these medium- and high-technology imports, respectively, accounted for 27 percent and 22 percent of South Africa's total merchandise imports (figure 1A.3), whereas low-technology imports averaged around 6 percent of total merchandise imports between 2001 and 2009. Overall, between 2001 and 2009, total imports in these three technology categories averaged about 58 percent of South Africa's total merchandise

Figure 1A.2   South Africa's Exports of Low-, Medium-, and High-Technology Goods, Dollar Value, 2001–09

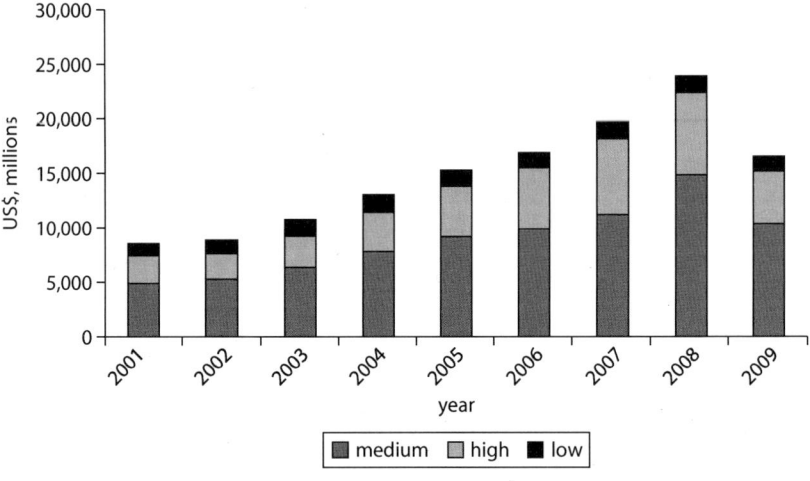

Source: SAIIA's calculations based on ITC data, 2010.

Figure 1A.3   South Africa's Imports of Low-, Medium-, and High-Technology Goods, 2001–09

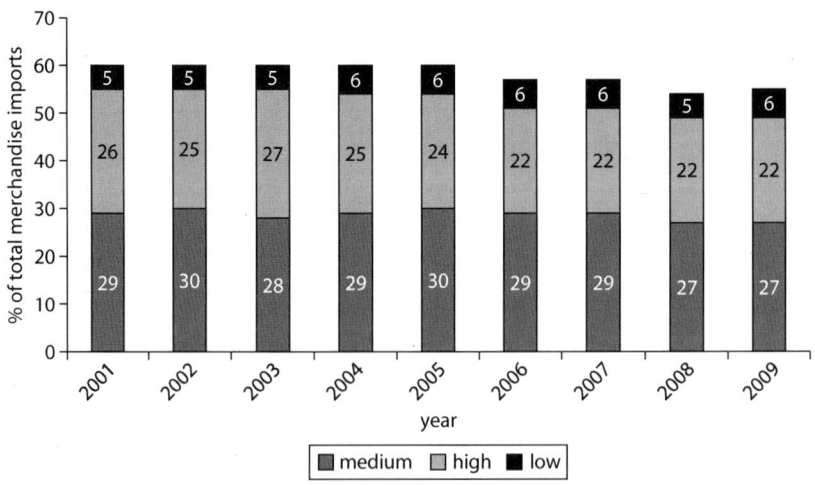

Source: SAIIA's calculations based on ITC data, 2010.

imports. Figure 1A.4 shows that South Africa's technology imports by value depict a similar trend as that of technology exports—growing since 2001, reaching a peak in 2008, and declining thereafter. Despite the growth, the proportional shares in absolute imports across technology groups remained stable.

Figure1A.4   South Africa's Imports of Low-, Medium-, and High-Technology Goods, Dollar Value, 2001–09

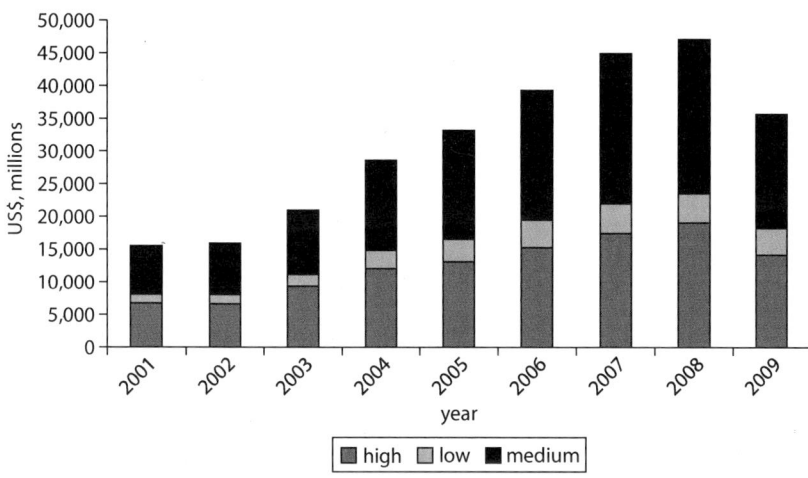

Source: SAIIA's calculations based on ITC data, 2010.

Structure of Mauritius's Trade in Technology-Intensive Goods

Technology exports. Mauritius's technology exports are dominated by low-technology goods, mainly clothing and textiles, which accounted for 39 percent of total merchandise exports in 2008, after declining from 61 percent in 2001 (see figure 1A.5). The cause of the decline is uncertain, but during the same period, Mauritius's total merchandise exports grew by 58 percent, while low-technology exports declined by 1.2 percent, characterized by a sharp decline in 2004–05.[11] This drop probably reflects the increased competition from Asian exporters of clothing and textiles after the phasing out of the MFA at the end of 2004. Subsequent recovery in terms of absolute values may reflect post-MFA consolidation and diversification into niche value-added clothing exports for the European market, and developments in the United States' African Growth and Opportunities Act, under which Mauritius was temporarily granted derogation from the third-country-sourcing rule of origin provision.

High- and medium-technology exports together accounted for about 10 percent of total merchandise exports in 2008 (figure 1A.5). As of 2008, approximately 40 percent of Mauritius's US$2.4 billion merchandise exports (figure 1A.6) have been attributed to low-technology exports (mainly apparel and clothing products). Sugar and

Figure 1A.5   Mauritius's Exports of Low-, Medium-, and High-Technology Goods, 2001–08

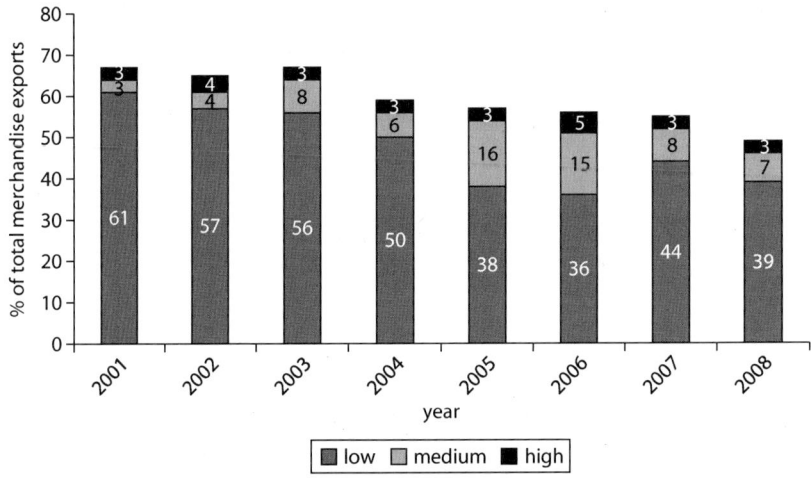

Source: SAIIA's calculations based on ITC data, 2010.

Figure 1A.6   Mauritius's Exports of Low-, Medium-, and High-Technology Goods, Dollar Value, 2001–08

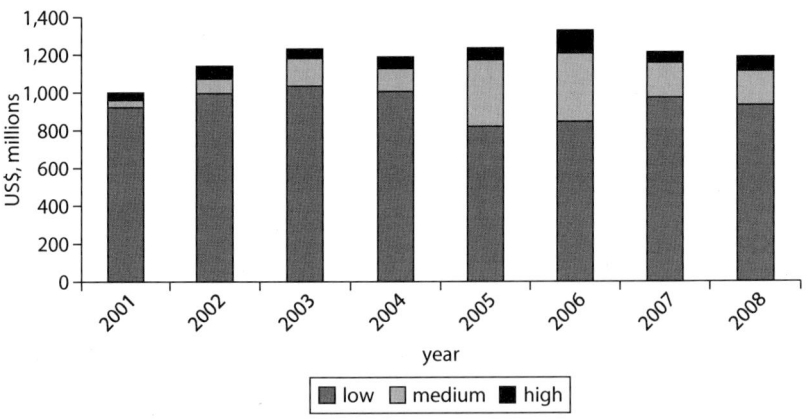

Source: SAIIA's calculations based on ITC data, 2010.

sugar confectionary (resource-based manufactures) constitute the second-largest segment of exports, accounting for approximately 11 percent of total exports. The remaining exports are fragmented across a broad range of products, including processed food, electrical machinery, and equipment. The share of high-technology exports was stable at about 3 percent from 2001 to 2008. Medium-technology exports have been erratic, with abrupt increases in 2005 and 2006 influenced by increased exports of machinery (electrical and electronic equipment), which accounted for 80 percent of Mauritius's medium-technology exports. However, in the textile sector, increasing competition from Asian markets has led Mauritius to cater to niche markets in Europe, leading to an emphasis on the quality of the textile.

High- and medium-technology exports in Mauritius remain very low, suggesting very little diversification of Mauritius's manufactured exports into higher-technology exports. However, as noted in the case of clothing exports (under Lall's 2000 classification, these are included in the low-technology category), this aggregate finding conceals some potentially significant sectoral diversification into relatively higher value-added activities.

Technology imports. Mauritius's technology imports are dominated by medium- and high-technology goods, which accounted for 20 percent and 11 percent, respectively, of total merchandise imports in 2008 (see figure 1A.7). During the same year, the share of low-technology imports accounted for 8 percent of total merchandise imports, a share that was relatively stable in the period under review. The value of Mauritius's technology imports is shown in figure 1A.8; the shares are

Figure 1A.7   Mauritius's Imports of Low-, Medium-, and High-Technology Goods, 2001–08

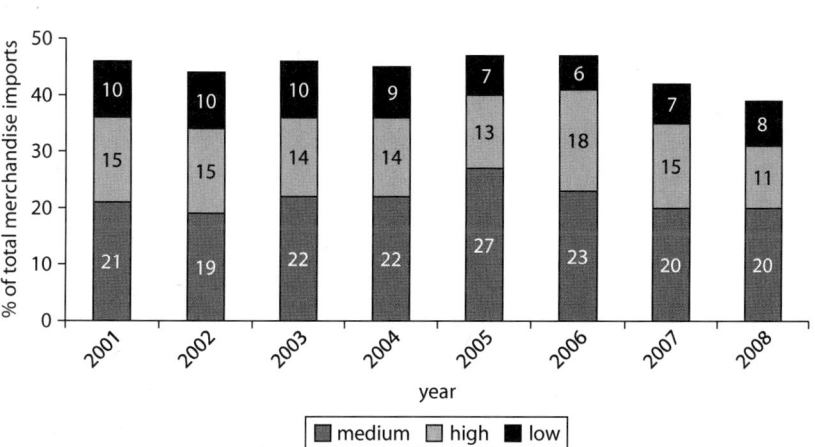

Source: SAIIA's calculations based on ITC data, 2010.

Figure 1A.8   Mauritius's Imports of Low-, Medium-, and High-Technology Goods, Dollar Value, 2001–08

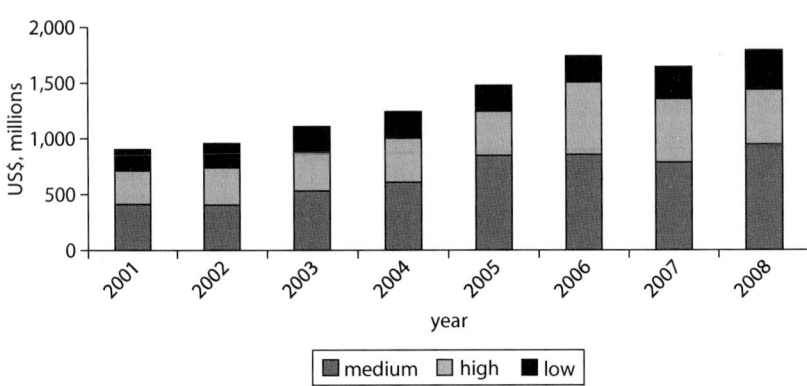

Source: SAIIA's calculations based on ITC data, 2010.

similar to those shown in figure 1A.7. Overall, 2005 exhibits a relative disjuncture from the normal pattern, particularly for medium-technology imports.

Low-technology imports have been relatively low and stable over the years. The analysis suggests that Mauritius relies on imports for its technology needs.

Structure of Namibia's Trade in Technology-Intensive Goods

Technology exports. Namibia's technology-intensive exports are very small as a proportion of total merchandise exports. Low-, medium-, and high-technology exports together accounted for just 9 percent of total merchandise exports in 2008. Medium-technology goods are the major category, which in 2008 stood at 5 percent of total merchandise exports (see figure 1A.9). Clearly, Namibia's exports have very low technology intensity; between 2001 and 2008, low-, medium-, and high-technology exports combined averaged about 11 percent of total merchandise exports. The same trends emerge when the total value of exports by technology category is considered (see figure 1A.10).

Technology imports. Namibia's technology imports are dominated by medium-technology goods (mainly transport equipment, machinery, and chemical products) and high-technology goods (mainly machinery and pharmaceutical products), which accounted for 29 percent

Figure 1A.9   Namibia's Exports of Low-, Medium-, and High-Technology Goods, 2001–08

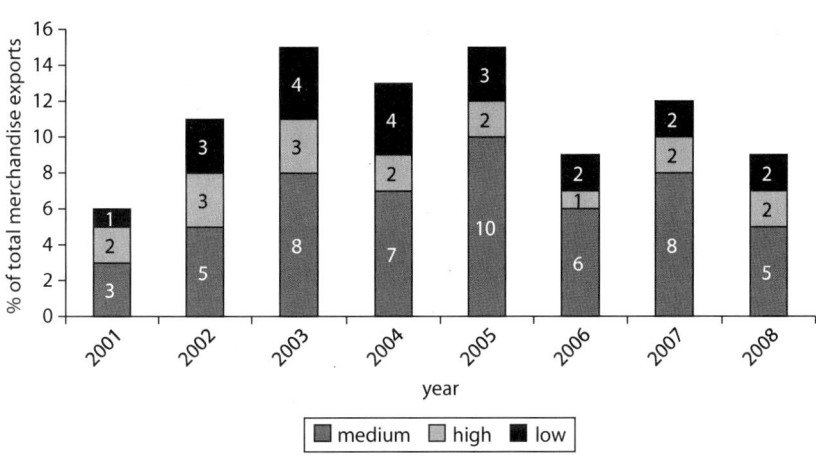

Source: SAIIA's calculations based on ITC data, 2010.

Figure 1A.10   Namibia's Exports of Low-, Medium-, and High-Technology Goods, Dollar Value, 2001–08

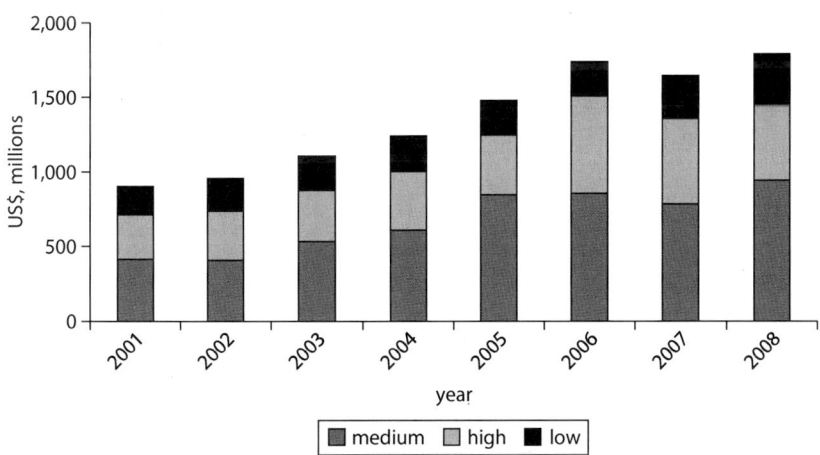

Source: SAIIA's calculations based on ITC data, 2010.

and 15 percent, respectively, of Namibia's total merchandise imports in 2008 (see figure 1A.11). In the same year, low-technology imports accounted for 10 percent of total merchandise imports. The shares of all technology categories have been stable since 2001. Figure 1A.12 shows that Namibia's low-, medium-, and high-technology imports

Figure 1A.11   Namibia's Imports of Low-, Medium-, and High-Technology Goods, 2001–08

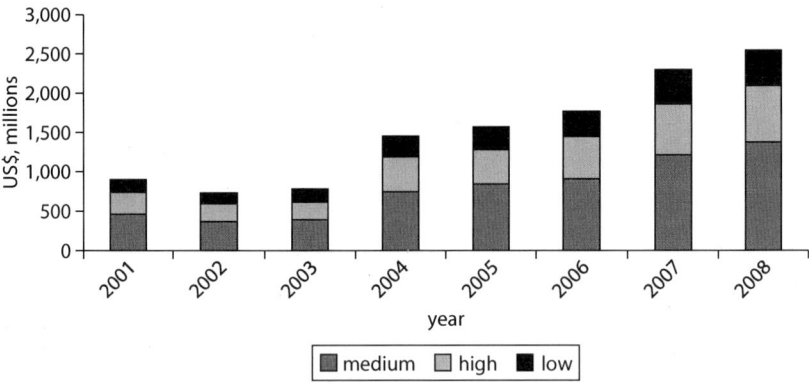

Source: SAIIA's calculations based on ITC data, 2010.

Figure 1A.12   Namibia's Imports of Low-, Medium-, and High-Technology Goods, Dollar Value, 2001–08

Source: SAIIA's calculations based on ITC data, 2010.

have been increasing since 2001. Nonetheless, the shares across technology categories have been stable.

Structure of Lesotho's Trade in Technology-Intensive Goods

Technology exports. Lesotho's technology exports are completely dominated by low-technology goods; clothing and textiles account for almost

100 percent of low-technology exports (see figure 1A.13). In 2003, low-technology manufactures accounted for 98 percent of Lesotho's total merchandise exports. However, the share of low-technology exports in total merchandise exports declined from 98 percent in 2003 to 56 percent in 2008. The same trends emerge when the total value of exports by technology category is considered (see figure 1A.14). This drop reflects the end of the MFA and increased competition from Asian exporters of clothing

Figure 1A.13   Lesotho's Exports of Low-, Medium-, and High-Technology Goods, 2001–08

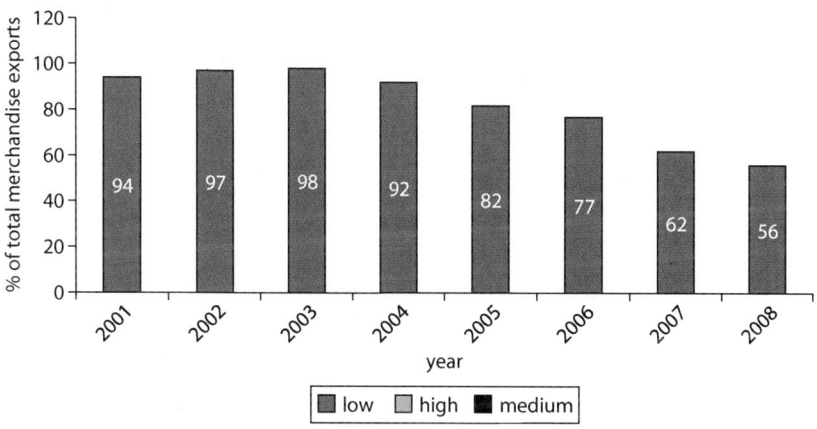

Source: SAIIA's calculations based on ITC data, 2010.

Figure 1A.14   Lesotho's Exports of Low-, Medium-, and High-Technology Goods, Dollar Value, 2001–08

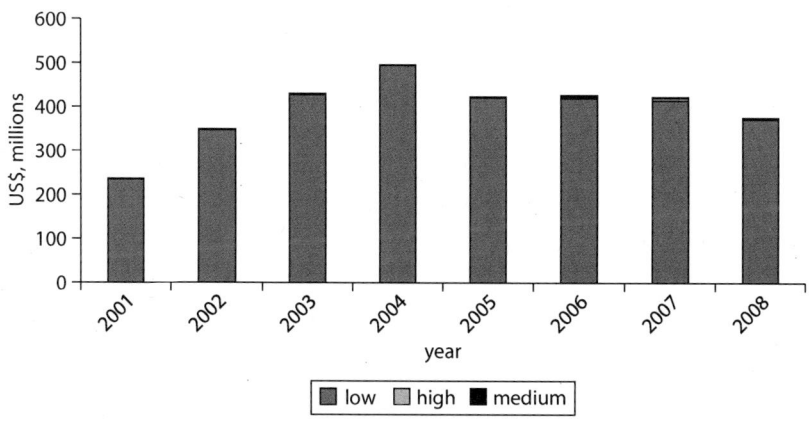

Source: SAIIA's calculations based on ITC data, 2010.

and textiles in lucrative markets, particularly the United States, and the fact that resource-based manufactures (not shown in the figure) are driven mainly by precious stones.

Technology imports. Lesotho's technology imports are dominated by low- and medium-technology goods (see figure 1A.15), with the latter becoming relatively more important toward the end of the period and the former declining in relative importance and absolute values (figure 1A.16). The share of medium-technology manufactures and the absolute values have been increasing since 2006, driven mainly by imports of electrical and electronic equipment in that year and subsequently. In 2008, the shares of low- and medium-technology imports in Lesotho's total merchandise imports were 35 percent and 33 percent, respectively. The share of high-technology imports has been erratic and accounted for 8 percent of total merchandise imports in 2008.

The structure of Lesotho's technology imports suggests that imports play a major role in the country's technological absorption. These imports are divided between low-technology imports (to support the clothing and textiles industry) and recent growth in medium-technology imports, particularly electrical and electronic machinery. High-technology imports were consistently marginal.

The following two tables provide an overview of the FDI flows in the four countries in recent years.

Figure 1A.15   Lesotho's Imports of Low-, Medium-, and High-Technology Goods, 2001–08

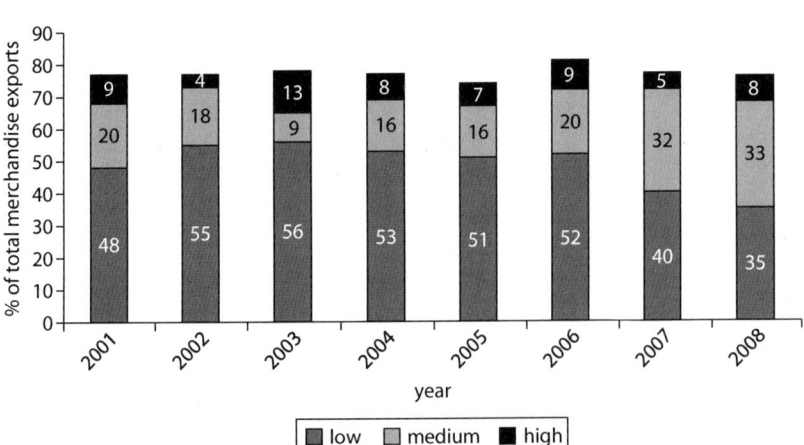

Source: SAIIA's calculations based on ITC data, 2010.

Figure 1A.16   Lesotho's Imports of Low-, Medium-, and High-Technology Goods, Dollar Value, 2001–08

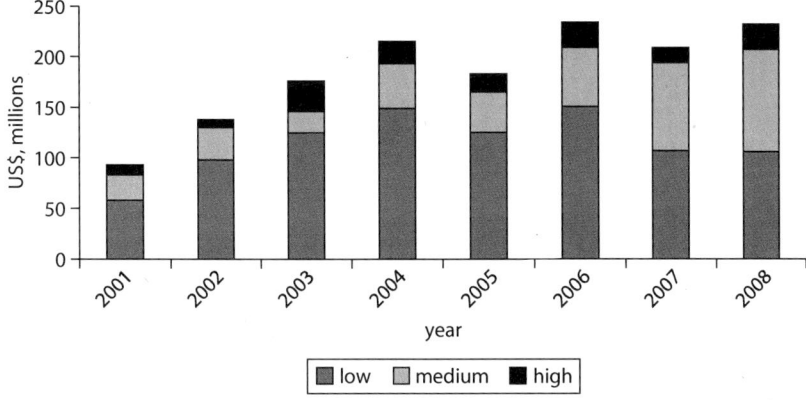

Source: SAIIA's calculations based on ITC data, 2010.

Table 1A.2   FDI Comparative Overview of Four Countries

Data	South Africa	Namibia	Lesotho	Mauritius
FDI inflows (US$, millions)				
2009	5,696	516	48	257
2008	9,006	720	56	383
2007	5,692	697	106	339
2006	−527	387	92	105
FDI stocks (US$, millions)				
2009	125,085	3,988	1,075	1,889
2008	68,007	3,472	934	1,632
2007	93,474	3,822	735	1,248
2006	87,782	2,786	629	910
FDI stocks (% GDP)				
2009	44	42	66	22
2008	24.6	39	58	18
2007	34	57	46	17
2006	35	44	44	14
Flows by country of origin,[a] 2007 (US$, millions)	No data	No data	• China: 0.5 (2005) • Germany: −2.5 (2004)	• United Kingdom: 88 • United States: 74 • Switzerland: 40 • Dubai: 40 • France: 37

(continued next page)

Table 1A.2   *(continued)*

Data	*South Africa*	*Namibia*	*Lesotho*	*Mauritius*
Stocks by country of origin, 2007 (US$, millions)	• United Kingdom: 74,620 • United States: 6,598 • Germany: 5,888 • Netherlands: 4,122 • Switzerland: 3,038	• United Kingdom: 125.0 • Iceland: 1.7 • Denmark: 1.1 (all 2006)	• South Africa: 34.3 • United States: 2.0 • China: 0.5 (all 2006)	No data
By country of origin, 2006 (% total)	• United Kingdom: 72.0 • United States: 6.1 • Germany: 5.6 • Japan: 2.4	No data	No data	• United Kingdom: 52.9 • United States: 6.1 • Switzerland: 8.1 • India: 2.2
Flows by sector, 2007 (US$, millions)	• Mining: 11,658 • Business services: 2,286 • Manufacturing and other secondary: 1,660 • Trade and hospitality: 1,650	No data	No data	• Tourism: 187 • Banking: 127 • Real estate: 32 • Manufacturing: 9
Stocks by sector, 2006 (US$, millions)	• Mining: 37,303 • Business services: 24,216 • Manufacturing and other secondary: 24,649 • Trade and hospitality: 2,410 • Transport and communications: 2,057	• Mining: 1,008 • Business services: 1,246 • Manufacturing and other secondary: 207 • Trade and hospitality: 371	No data	No data
Stocks by sector, 2006 (% total)	• Mining: 34.4 • Manufacturing and other secondary: 27.8 • Business services: 32.2	Mining: 76.6 Business services: 18.1 Manufacturing and other secondary: 5.3	No data	Business services: 48.6 Trade and hospitality: 36.1

(continued next page)

Table 1A.2   *(continued)*

Data	South Africa	Namibia	Lesotho	Mauritius
Stocks by sector and country of origin, 2006 (US$, millions)	Mining: • EU: 12,860 • Developing countries: 461 • North America: 161 Business services: • EU: 17,721 • Developing countries: 26 • North America: 968 Manufacturing and other secondary: • EU: 10,589 • Developing countries: 839 • North America: 1,558 Trade and hospitality: • EU: 1418 • Developing countries: 551 • North America: 231 Transport and communications: • EU: 489 • Developing countries: 460 • North America: 362	No data	No data	No data

Sources: Data from Bank of Mauritius; South African Reserve Bank, 2010; UNCTAD 2010.
a. Only the largest investors and sectors are shown.

Table 1A.3   Stock and Flow Contrast with Comparator Countries, 2009

Country	Inflows (US$, millions)	Percentage of gross fixed capital formation	Stocks (US$, millions)	Percentage of GDP
Jamaica	1,062	41	11,166	90
Lesotho	48	9	1.075	66
Mauritius	257	12	1,889	22
Namibia	516	25	3,988	42
Poland	11,395	13	182,799	43
South Africa	5,696	9	125,085	44
Vietnam	4,500	13	52,825	52

Source: UNCTAD 2010.

Notes

1. This perspective has a long pedigree going back to Gerschenkron (1962) and before that to Veblen (1915).

2. Internationally comparable data are not available for other southern African countries.

3. The technological content of exports and imports in the four countries was reviewed using the technology-based product classification developed by Lall (2000). Lall's approach is based on the second revision of the Standard International Trade Classification of commodities in which products are classified into primary products, resource-based manufactures, low-technology manufactures, medium-technology manufactures, and high-technology manufactures (see table 1A.1 in the annex). However, the analysis throughout excludes special transactions, primary products, and resource-based products because these three categories add little from a technological content perspective. Primary products and resource-based manufactures tend to be unskilled labor and scale intensive, and skill requirements tend to rise with the degree of technological complexity. (For a complete description of each technology category, please refer to Lall [2000].) Consequently, three technology categories (low, medium, and high) are expressed as percentages of total merchandise (not just manufactured) exports, in conformity with Lall's approach.

4. Although the sector as a whole is classified as low technology, Mauritian garment exports are catering to niche markets that require some technology for sophisticated design and material handling, for example.

5. The United Nations Conference on Trade and Development lists FDI performance and potential indexes. In 2009, Namibia ranked 34th of 141 countries for performance and 89th for potential. In other words, it is attracting more FDI than expected by its investment climate. South Africa was ranked 79th for performance and 72nd for potential. It is slightly below its potential.

6. Data are from the World Bank's *World Development Indicators*, http://databank.worldbank.org.

7. Dolby Laboratories, which specializes in innovative sound technologies, makes about 78 percent of its revenues from licensing, 14 percent from product sales, and 7 percent from services. Its products are used in entertainment audio systems (such as headphones, computers, DVDs, and movies) that are produced by other firms.

8. Lederman (2009), using firm-level data, finds evidence that licensing is significantly and positively correlated with the probability that a firm innovates, and so is a firm's expenditure on R&D. The diaspora and other networks also represent an important channel for technology diffusion. Not only can the diaspora serve as a significant source of skills, entrepreneurial ability, and business and marketing expertise, but it can also contribute by strengthening trade and investment linkages (see World Bank 2008).

9. The main organization presiding over the development of the Community Innovation Surveys is Eurostat, a Directorate-General of the European Commission based in Luxembourg. It, in turn, works with national statistical agencies for European Union member countries carrying out the surveys. One of the key efforts by the organization is to harmonize the questionnaires to the greatest degree possible.

10. For firms that did not respond, a nonresponse follow-up survey was conducted for 15 percent of nonrespondents. The purpose of the nonresponse survey was to determine the extent to which nonrespondents are less or more innovative than respondents. The HSRC concluded that the qualitative data in the survey were generally representative of firms (Blankley and Moses 2009, 7) and therefore appropriate for analysis.

11. Authors' calculations from International Trade Commission data.

References

Aghion, Philippe, Nick Bloom, Richard Blundell, Rachel Griffith, and Peter Howitt. 2005. "Competition and Innovation: An Inverted U Relationship." *Quarterly Journal of Economics* 120 (2): 701–28.

Aghion, Philippe, Matias Braun, and Johannes Fedderke. 2008. "Competition and Productivity Growth in South Africa." *Economics of Transition* 16 (4): 741–68.

Blankley, William, and Cheryl Moses. 2009. *Main Results of the South African Innovation Survey 2005*. Cape Town: Human Sciences Research Council Press.

Blomström, Magnus, and Ari Kokko. 2003. "The Economics of Foreign Direct Investment Incentives." NBER Working Paper 9489, National Bureau of Economic Research, Cambridge, MA.

Chandra, Vandana, ed. 2006. *Technology, Adaptation, and Exports: How Some Developing Countries Got It Right*. Washington, DC: World Bank.

Coe, David, and Elhanan Helpman. 1995. "International R&D Spillovers." *European Economic Review* 39 (5): 859–87.

Cohen, Wesley, and Daniel Levinthal. 1989. "Innovation and Learning: The Two Faces of R&D." *Economic Journal* 99 (397): 569–96.

Commission on Growth and Development. 2008. *The Growth Report: Strategies for Sustained Growth and Inclusive Development.* Washington, DC: World Bank on behalf of the Commission on Growth and Development.

De Ferranti, David, Guillermo Perry, Indermit Gill, J. Luis Guasch, William Maloney, Carolina Sánchez-Páramo, and Norbert Schady. 2003. *Closing the Gap in Education and Technology.* Washington, DC: World Bank.

Dutz, Mark. 2007. *Unleashing India's Innovation Potential.* Washington, DC: World Bank.

Eaton, Jonathan, and Sanuel Kortum. 1996. "Trade in Ideas: Patenting and Productivity in the OECD." *Journal of International Economics* 40 (3–4): 251–78. Cited in Saggi, Kamal. 2002. "Trade, Foreign Direct Investment, and Technology Transfer: A Survey." *World Bank Research Observer* 17 (2): 191–235.

Enterprise Surveys (database). World Bank, Washington, DC. https://www.enterprisesurveys.org.

Freeman, Chris. 1988. "Japan: A New National Innovation System?" In *Technical Change and Economic Theory*, ed. G. Dosi, C. Freeman, R. R. Nelson, G. Silverberg, and L. Soete, 330–48. London: Pinter.

Gerschenkron, Alexander. 1962. *Economic Backwardness in Historical Perspective: A Book of Essays.* Cambridge, MA: Harvard University Press.

Goldberg, Itzhak, Lee Branstetter, John G. Goddard, and Smita Kuriakose. 2008. "Globalization and Technology Absorption in Europe and Central Asia: The Role of Trade, FDI, and Cross-Border Knowledge Flows." Working Paper 150, World Bank, Washington, DC.

Goldberg, Itzhak, Manuel Trajtenberg, Adam Jaffe, Thomas Muller, Julie Sunderland, and Enrique Blanco Armas. 2006. "Public Financial Support for Commercial Innovation." Regional Working Paper Series sponsored by the Chief Economist's Office in the Europe and Central Asia Region, World Bank, Washington, DC.

Griffith, Rachel, Stephen Redding, and John Van Reenen. 2004. "Mapping the Two Faces of R&D: Productivity Growth in a Panel of OECD Industries." *Review of Economics and Statistics* 86 (4): 883–95.

Hausmann, Ricardo, and Dani Rodrik. 2003. "Economic Development as Self-Discovery." *Journal of Development Economics* 72 (2): 602–33.

Hoekman, Bernard, and Beata Javorcik. 2006. "Lessons from Empirical Research on Technology Diffusion through Trade and Foreign Direct Investment." In

Global Integration and Technology Transfer, ed. Bernard Hoekman and Beata Javorcik, 1–28. Washington, DC: World Bank; New York and Houndmills, Basingstoke, Hampshire: Palgrave Macmillan.

HSRC (Human Sciences Research Council). 2005. *South Africa Innovation Survey 2005.* Department of Science and Technology, Pretoria, South Africa.

Kaplan, David. 2008. "Science and Technology Policy in South Africa: Past Performance and Proposals for the Future." *Science, Technology and Society* 13 (1): 95–122.

Keller, Wolfgang. 2004. "Trade and the Transmission of Technology." *Journal of Economic Growth* 7 (1): 5–24.

Khan, Mushtaq. 2009. "Learning, Technology Acquisition and Governance Challenges in Developing Countries." U.K. Department for International Development–funded research document.

Kinoshita, Yuko. 1998. "Technology Spillovers through Foreign Direct Investment." William Davidson Institute Working Paper No. 221, University of Michigan, Ann Arbor, MI.

Lall, Sanjaya. 2000. "The Technological Structure and Performance of Developing Country Exports, 1985–1998." QEH Working Paper 44, Queen Elizabeth House, University of Oxford, U.K.

Lederman, Daniel. 2009. "The Business of Product Innovation: International Empirical Evidence." Policy Research Working Paper 4840, World Bank, Washington, DC.

———. 2010. "An International Multi-level Analysis of Product Innovation." *Journal of International Business Studies* 41: 606–19.

Lin, Justin, and Celestin Monga. 2010. "Growth Identification and Facilitation: The Role of the State in the Dynamics of Structural Change." Policy Research Working Paper 5313, World Bank, Washington, DC.

Lundvall, Bengt-Åke. 1992. *National Innovation Systems: Towards a Theory of Innovation and Interactive Learning.* London: Pinter.

Mengistae, Taye. 2011. "Are South African Wages Too High or Growing Too Fast?: A Comparison of Manufacturing Pay and Productivity in Selected Middle Income Economies" (Draft). Background paper for ongoing study on economic diversification in South Africa, World Bank, Washington, DC.

Nelson, Richard, ed. 1993. *National Innovation System: A Comparative Analysis.* Oxford: Oxford University Press.

Pack, Howard, and Kamal Saggi. 2001. "Vertical Technology Transfer via International Outsourcing." *Journal of Development Economics* 65 (2): 389–415.

Pack, Howard, and Larry Westphal. 1986. "Industrial Strategy and Technological Change: Theory versus Reality." *Journal of Development Economics* 22: 87–128.

Quantec (database). 2011. http://www.quantec.co.za.

Racine, Jean Louis, Itzhak Goldberg, John G. Goddard, Smita Kuriakose, and Natasha Kapil. 2009. "Restructuring of Research and Development Institutes in Europe and Central Asia" (Draft April 13). Europe and Central Asia, Private and Financial Sector Development Department, World Bank, Washington, DC. http://siteresources.worldbank.org/INTECAREGTOPKNOECO/Resources/ECAKE3_April_13_09_Exec_Sum.pdf.

Rodrik, Dani. 2004. "Industrial Policy for the Twenty-First Century." Background paper prepared for UNIDO, John F. Kennedy School of Government, Harvard University, Cambridge, MA.

———. 2008. "Normalizing Industrial Policy." Working Paper 3, Commission on Growth and Development, World Bank, Washington, DC.

Saggi, Kamal. 2002. "Trade, Foreign Direct Investment, and International Technology Transfer: A Survey." *World Bank Research Observer* 17 (2): 191–235.

Sanders, Mark, and Bas ter Weel. 2000. "Skill-Biased Technical Change: Theoretical Concepts, Empirical Problems and a Survey of the Evidence." Draft paper presented at the DRUID Conference, Copenhagen, Denmark, January 6–8. http://www.merit.unu.edu/publications/rmpdf/2000/rm2000-012.pdf.

Schiff, Maurice, and Yanling Wang. 2006. "North-South and South-South Trade Related Technology Diffusion: An Industry Level Analysis of Direct and Indirect Effects." *Canadian Journal of Economics* 39 (3): 831–44.

UNCTAD (United Nations Conference on Trade and Development). 1995. *World Investment Report: Transnational Corporations and Competitiveness.* New York and Geneva: United Nations.

———. 2010. *World Investment Report: Investing in a Low-Carbon Economy.* New York and Geneva: United Nations.

UNECA (United Nations Economic Commission for Africa). 2010. *A Technological Resurgence? Africa in the Global Flows of Technology.* Addis Ababa, Ethiopia: United Nations.

UNIDO (United Nations Industrial Development Organization). 2009. *Industrial Development Report 2009: Breaking In and Moving Up—New Industrial Challenges for the Bottom Billion and the Middle-Income Countries.* Vienna: UNIDO.

Veblen, Thorstein. (1915) 2003. *Imperial Germany and the Industrial Revolution.* Kitchener, Ontario, Canada: Batoche Books.

Wang, Miao. 2009. "Manufacturing FDI and Economic Growth: Evidence from Asian Economies." *Applied Econometrics* 41 (8): 991–1002.

WEF (World Economic Forum). 2010. *Global Competitiveness Report 2010–11.* Geneva, Switzerland.

World Bank. 2004. *World Development Report 2004: Making Services Work for Poor People*. Washington, DC: World Bank.

———. 2008. *Global Economic Prospects: Technology Diffusion in the Developing World*. Washington, DC: World Bank.

———. 2009. "Productivity and the Investment Climate: Private Enterprise Survey: Mauritius" (Draft December). World Bank, Washington, DC.

———. 2010. *Stepping Up Skills: For More Jobs and Higher Productivity*. Washington, DC: World Bank.

———. 2011a. "Closing the Skills and Technology Gaps in South Africa" (Draft May). World Bank, Washington, DC.

———. 2011b. "Skills and Technology Absorption in Mauritius" (Draft May). World Bank, Washington, DC.

Xu, Bin, and Jianmao Wang. 1999. "Capital Trade Goods and R&D Spillovers in the OECD." *Canadian Journal of Economics* 32: 1258–74.

Yusuf, Shahid, and Kaoru Nabeshima. 2011. "Some Small Countries Do It Better: Rapid Growth and Its Causes in Singapore, Finland, and Ireland." World Bank, Washington, DC.

Zhang, Chunlin, Douglas Zhihua Zeng, William Peter Mako, and James Seward. 2009. *Promoting Enterprise-Led Innovation in China*. Directions in Development Series. Washington, DC: IBRD/World Bank.

CHAPTER 2

Channels of and Constraints to Technology Absorption

Smita Kuriakose, David Kaplan, and Krista Tuomi

This chapter summarizes findings from the case study interviews conducted in key exporting sectors in the four countries under review in this book, supplemented and complemented by statistical data drawn principally from Enterprise Surveys (ESs) but also, where available, by other industry studies. The basic survey instrument used has been attached in annex 2B. Specific modifications were made for each country. The aim of the analyses is to understand how technology is transferred and absorbed in southern African firms, what factors constrain the deepening of technological capabilities, how firms deal with these constraints, and to what extent firms find government policies helpful in overcoming these constraints. Evidence from the case interviews conducted and from the ESs supports the findings of Van Biesebroeck (2005),[1] suggesting that a relationship exists between technology, productivity, and exports. This chapter explores the complex relationship between technology and productivity, on the one hand, and exporting, foreign ownership, or both, on the other.

The authors gratefully acknowledge background papers from Justin Barnes, Reza Daniels, Peter Draper, Mike Morris, and Eric Wood.

South Africa: Auto Component Sector

The automotive sector has grown rapidly in South Africa, constituting 5.5 percent of gross domestic product (GDP) in the early 1990s and 7.5 percent in 2008. The sector's share in exports has risen even more dramatically, from 4 percent in 1995 to 14 percent in 2007. The auto component sector currently has a turnover of some R 75 billion (approximately US$10.8 billion) with 75,000 employees. Automotive components currently constitute approximately 55 percent of total automotive exports. Although components are exported to a very wide range of markets, the European Union countries are the dominant destination, with Germany responsible for one-third of all South African component exports (tables 2A.1 and 2A.2 in annex 2A).

Though a wide range of components are exported, the export basket is dominated by catalytic converters, which have a rather low-technology content and constitute more than half of all component exports (see table 2A.3 in annex 2A). In the period 2004–08, catalytic converters were responsible for more than three-quarters of the growth in component exports.[2] Furthermore, if the component exports excluding catalytic converters are deflated by the effective exchange rate of the rand against South Africa's trading partners, the increase since 2004 has been marginal.[3]

The auto component sector has benefited greatly from the Motor Industry Development Programme (MIDP). Introduced in 1995, it was an export-import complementation scheme whereby duty-free importation was provided conditional on export performance. The Automotive Production and Development Programme announced in September 2008 is the successor to the MIDP. The program is composed of four elements, which include tariffs of 25 percent for vehicles and 20 percent for components; a local assembly allowance whereby assemblers with plant volume of at least 50,000 units can import a percentage of their components duty free; and duty credits (the local assembly allowance comes in the form of duty credits) issued to vehicle assemblers based on 18–20 percent of the value of the motor vehicles produced from 2013. In addition, manufacturers are to receive support to encourage the use of local components.

Technology Absorption in the Auto Component Sector

Forty percent of total production is directly exported (NAACAM 2010). In the two types of exports, component firms either (a) supply the auto assemblers (original equipment manufacturers, or OEMs) who supply another country's OEM, or (b) export to the aftermarket. Both "customers" require that these products meet the very demanding technical specifications that are set by the OEMs. As with auto component producers everywhere, South African component suppliers are located in a global system that tightly regulates product specifications. South African producers must meet these product standards, which require them to employ the most up-to-date production techniques.

Technology development is strongly focused on the production process. In only a few sector areas where the South African market has particularized characteristics, such as alarms and antitheft devices, have local firms developed some unique products. For the industry as a whole, the emphasis of technological development is very much on the production process and, to a limited extent, on minor product adaptation, generally for the domestic market.

Expenditure on research and development (R&D) in the South African auto component industry is very low. Only one-third of auto component firms in the South African Automotive Benchmarking Club (SAABC) registered any R&D expenditures.[4] Moreover, R&D expenditures have been declining significantly (Gastrow 2008, 9). For those firms in the SAABC, R&D as a share of turnover declined from 1.62 percent in 2005 to 1.46 percent in 2009 (SAABC, personal correspondence with author). The firms in this sector have benefited from the MIDP, which has specific incentives for exporters. The particular programs in place for technology absorption or innovation have not been accessed by the firms interviewed because they either lack knowledge of the schemes' existence or perceive that these schemes are onerous to access.

For most auto component products, the minimum efficient scale required for each product is generally much larger than that provided by domestic demand. This situation, complemented by export incentives, leads firms to increase their exports to realize scale economies of larger production runs. Thus, many South African auto component firms are characterized by a wide product range to supply service markets abroad. Fierce competition and the growing penetration of products from Asia, notably in the lower end of the product market, have pressed producers to attempt to enter new, higher-value product markets. The large component producer interviewed had increased exports by significantly expanding its product range in the wake of the current recession, which had significantly lowered domestic demand for its products. A greater product range, however, adds to the complexity of the business and requires, in turn, new technological competencies to develop new and frequently more complex products.

One of the firms interviewed had invested in the latest technological innovation in glass technology to maintain its stronghold in the export market and continue to increase its export markets. This particular auto component producer had to acquire competencies to build more complex models, thus increasing demand for more advanced technology. Hence, the growth in exports has led to an increased demand for technological developments in the production process.

Channels of technology transfer. Thus, although the main driver to export is to ensure volume in a scale-intensive industry, increasing exports frequently entails introducing new and more diversified products. This diversification, in turn, affects the technological competencies of South African auto component exporters.

Acquisition of machinery and equipment. The 2003 ES data for South Africa had 20 firms from the auto component sector. Sixty-five percent of firms exported more than 10 percent of production, and 35 percent of firms were foreign owned. The ES data also show that acquisition of machinery and equipment is the most important source of technology transfer among both domestic and foreign-owned firms in the sector. The top technology channels for foreign firms in the sector in the sample from the 2003 ES were as follows: embodied in machinery, 25 percent; developed with a supplier, 13 percent; developed in house, 13 percent; and obtained through hiring of personnel, 13 percent. The top three channels for domestic firms in the sector were embodied in machinery, 25 percent; obtained through domestic licensing, 17 percent; and developed in house, 19 percent.

Another very recent survey corroborates that the acquisition of machinery is the most important innovation-related transaction for both foreign and local firms (Buys 2010, 2620). Auto component manufacturers source their technology from employees, international suppliers, and clients. Universities and research institutions are not important sources of knowledge in this industry. No significant differences were seen between domestic and foreign firms in terms of personnel (number of employees, staff composition, and level of education); systems of production organization; and types of products. However, the standard of machinery and equipment used by foreign firms was ahead of the average for the industry by three to four years, and foreign firms were more likely to engage in R&D and to employ more personnel in R&D than local firms (Buys 2010, 2615–16).

Technology agreements with global players. To ensure that they meet the demanding product standards, South African auto component firms often have technology agreements with global players. As a condition of purchase of the firm's products, the OEM that the South African firm supplies will often require that such an agreement is in place. One of the firms interviewed made cables using a technical agreement with one of the leading producers that supplied the same OEMs. When the company decided to move into producing mirrors, which was a new product line, a condition set by the OEM was that the South African firm must engage in a technical agreement with a certain major global player in that area. A plant visit showed that the South African firm used production methods obtained from the global player. The global player provided detailed instructions on the production process and also provided much of the equipment used. This process ensured direct and supervised technology transfer from the global player to the local South African firm.

South African auto component firms commonly obtain licenses for technology. Just over half the firms in the SAABC database for whom data are available paid royalties. Royalty payments in 2009 were 2.23 percent of applicable sales. In the ES sample of firms, foreign firms were far more likely to license technology compared with domestic firms. In fact, all the foreign firms in the sample licensed foreign technology, compared with only 17 percent of the domestic firms in the sector.

Collaboration with suppliers of raw materials. One of the firms interviewed produced hot metal forgings for first-tier OEMs and therefore has a competitive advantage in metallurgy. It worked closely with the major steel supplier to produce a unique steel grade that increased production efficiency. This collaboration has been vital to the firm's ability to differentiate its products and remain competitive.

Constraints on enhancing technological competencies. South Africa has an auto assembly industry that dates back 90 years, and it has some long-established companies in auto component production. Access to technology is not a major problem for South African auto component producers, many of whom have technical agreements with leading global players in their fields. In other areas, such as foundries, the technology is widely diffused. Machinery can be freely imported and attracts no tariff duties. The emphasis is currently on enhancing technology to compete effectively with low-cost producers from Asia.

Skill shortages and high labor costs. Most of the firms interviewed identified the shortage of skills as the major factor currently restraining technology development. This finding is true for high-level skills as well as
technical and artisanal skills such as drafting. Firms are of the view that
the situation is getting worse, and particularly that the higher-level skills
are even more difficult to access. The 2003 ES data corroborate this
finding for the industry as a whole, where the skill shortage was the top
concern among firms surveyed. Splitting the sample by exporter status
and ownership status also yielded similar results, with skill shortage
coming out as the top constraint (figures 2.1 and 2.2). This sector also
suffers from a high cost structure, to which the high labor costs are a
major contributor.

A recent study comparing the auto assembly sectors in South Africa
and Thailand concluded that Thai exporters enjoyed a 13.5 percent
advantage on cost of sales over a South African exporter, 11.7 percentage

**Figure 2.1   Top Constraints: Foreign-Owned versus Domestic Firms, South African
Auto Component Sector**

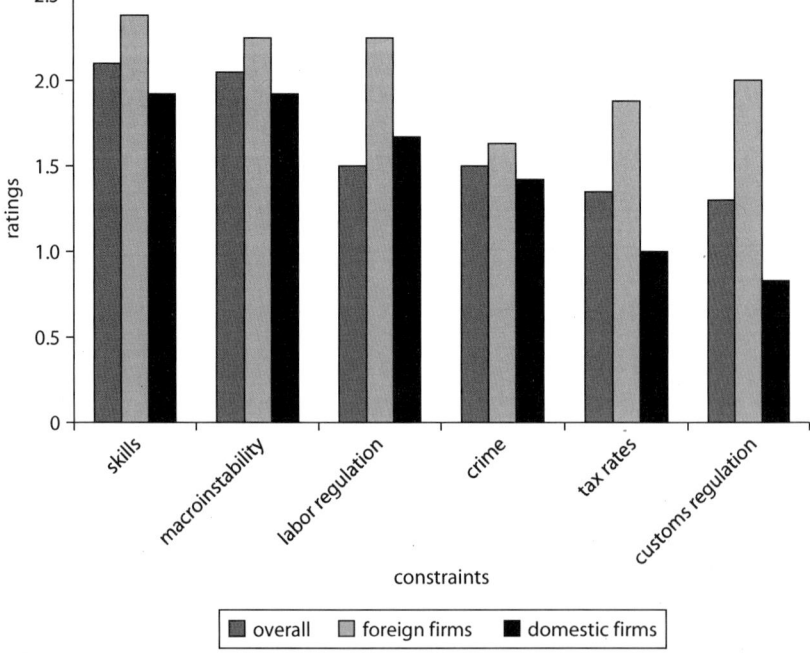

Source: Enterprise Survey (database), 2003 data.

Figure 2.2   Top Constraints: Exporters versus Nonexporters, South African Auto Component Sector

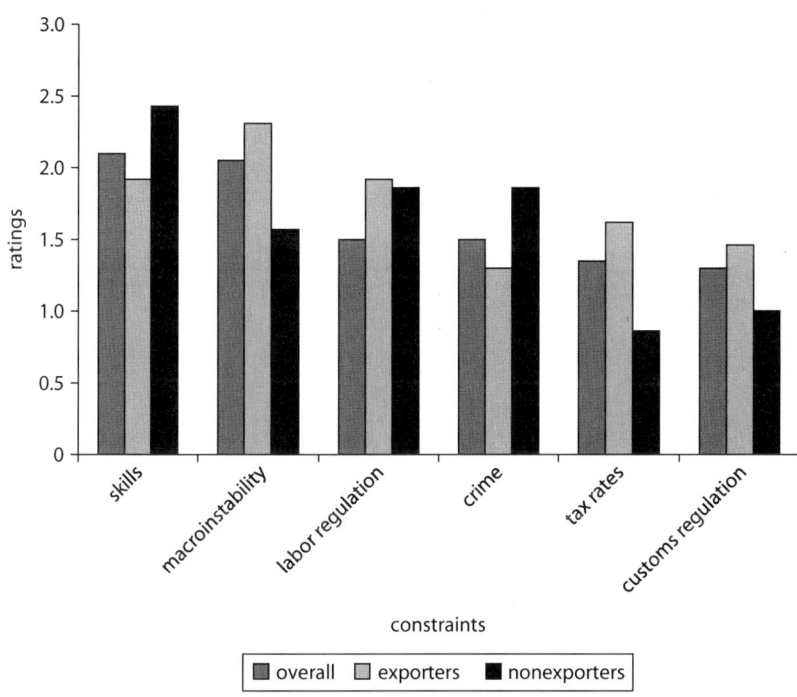

Source: Enterprise Survey (database), 2003 data.

points of which was caused by the higher cost of labor in South Africa. South African labor was more expensive across all categories, with the largest differential between the two countries at the artisanal level.[5] Firms respond by training, which is extensive in this sector.[6] The finding is also corroborated among the firms surveyed in the ES, with 75 percent of firms in the sector claiming to offer training. Some firms seek to provide training in state-of-the-art facilities through explicit technology agreements with leading global players. However, few firms in the sector as a whole make use of the Sector Education and Training Authority (SETA) system. Only one-fifth of the auto component firms accessed the SETA system; 13 percent of the firms were local, and only 6 percent were foreign (Stijger and Steyn 2010, 2630).

Higher wage costs are causing firms to rationalize their workforce and to adopt more capital-intensive production methods, and the firms pay skilled workers higher wages to attract them and keep them in the firm.

However, labor regulations are said to be making any reductions in the workforce more difficult.

Lack of effective industry-research links. Component-producing firms have very little relation with tertiary education institutions in terms of research. Somewhat stronger relations are seen with the Council for Scientific and Industrial Research (CSIR), which provides technical support and advice as well as testing services. A recent survey similarly concluded that very few firms have links with universities or research centers, whether local or international. "Local, domestic or international universities or research centers are not important sources of technology and knowledge for firms in this industry. Very few firms consider collaboration with universities or research centers important for activities such as training and R&D, and although foreign firms have a somewhat higher propensity for such collaboration than domestic firms, the difference is not statistically significant" (Buys 2010, 2620). For the firm producing automotive glass, technical competencies in glass technology have decreased in South Africa. The links with the global partner for the firm are vital in this regard. The firm is moving from a loose consultative arrangement with the global partner to a more formal arrangement that would enable the firm to access the training facilities of the global partner.

Summary

The auto component sector is characterized primarily by technology development that is strongly focused on the production process. Though new product innovation is not widespread, compliance with the high standards of the OEMs requires auto component sector firms to invest in technology. The main channels of technology transfer that are specific to this sector are the technology agreements that these firms have with their global partners. The lack of skilled personnel and inadequate industry-research links increase the importance of these global partners, which are increasingly becoming a vital source of training and technical know-how in light of the declining local skill competencies.

South Africa: Capital Goods Sector

Capital goods cover a wide variety of products used in mining, agriculture and agribusiness, building and construction, industrial processing, and utilities. The capital goods industry in South Africa has developed in response to local demand, particularly from the mining sector. Despite a lack of cohesive industrial policy with respect to capital goods, successive governments have initiated policies that have encouraged further development in the sector. Currently, the government has an active industrial policy, and capital equipment is one of the key areas identified for policy support.

Capital equipment exports are currently R 65 billion annually (about US$9.35 billion), representing some 6 percent of South Africa's total commodity exports. Apart from a dip during the 2008 recession, particularly in the export of mining equipment, this share has been growing. African countries, notably in the southern Africa region, are the major markets for South African exports of capital goods. Eight of the top 10 destination countries are in Africa, and all except Nigeria are located in southern Africa (see table 2A.4 in annex 2A). Of capital equipment exports, 37 percent are destined for five countries in the immediate region—Angola, the Democratic Republic of Congo, Mozambique, Zambia, and Zimbabwe. Of the top five export destination countries in 2007, only one, the United States, was outside the region.[7]

Mining equipment is the most dynamic part of the capital goods sector and constitutes more than 50 percent of total capital goods exports. Thirty-eight percent of mining equipment exports is destined for Sub-Saharan Africa, with the other major destination being Western Europe (tables 2A.5 and 2A.6 in annex 2A). Most of South Africa's exports to Sub-Saharan Africa are to countries in the region. Mining equipment exports have particularly high local value-added content. The South African Capital Equipment Export Council (SACEEC) estimates local content for mining equipment at approximately 90 percent. This level is significantly more than for other high- and medium-technology South African exports.

Technology Absorption in the Capital Equipment Sector

At the product level, South Africa is a world leader in a number of mining equipment products. These include spirals for coal washing and equipment for water pumping, hydropower, tracked mining, underground locomotives, ventilation, shaft sinking, and turnkey new mine design, among others. South African expertise is particularly advanced and at the global frontier in deep mining and associated competencies. South Africa is much weaker in so-called yellow-metal areas, such as mining vehicles, where scale economies are critical and where large multinational corporations (MNCs) dominate. Outside of mining, South Africa has leading global products in a number of other areas, such as transport and haulage equipment and processing equipment. In many cases, the mining sector provided the initial source of demand, and successful domestic firms then branched out into other areas.

South Africa has globally leading technologies in mining and mining-related activities and has a significant number of high-quality patents comparable, for example, to Australia. Firms in mining equipment develop much of their intellectual property locally. Outside of mining and related areas, however, South Africa has few patents, and producers tend to be technological followers rather than leaders.

A number of leading mining MNCs have subsidiaries in South Africa, where they undertake considerable design and development work. A significant number of South African firms, particularly producers of mining equipment, compete effectively with global MNCs in the domestic market as well as in other countries and regions. Although many of these companies are large, some are medium and small firms.

Mining is highly location specific, with varying geological and other conditions, which results in the development of "applied competencies," generally onsite. In South Africa, local conditions are highly specific and demanding. This is most evident with respect to gold mining, which has to be mined in hard rock at levels deeper than is necessary anywhere else. Coal and platinum also have distinctive properties. Both minerals occur in a form that is difficult to exploit. South Africa was able to develop coal and platinum metallurgy that allowed what were unprofitable deposits to be profitably mined. Many spin-off capacities also had significant implications for the development of mining equipment. For example, South African coal had to be washed to remove impurities. This necessity led to extensive development in spirals for washing, in which South Africa has major exports and indeed leads the world, and which is now being applied in a number of new areas, such as tar.

Many local capital goods firms have been suppliers to local mining operations over a very long period, thereby establishing close working relationships and trust. These local suppliers of mining equipment exports followed South African outward foreign direct investment (FDI) in mining when South African mining houses extended their operations overseas. Many major South African mining houses have, since 1994, undertaken significant expansions abroad, notably in Africa, creating considerable export opportunities for suppliers of mining equipment and specialist services. Local companies are using their mining-related technological expertise to develop products and services that are destined for other uses, in ventilation and hydraulic pumps, for example. This lateral movement is not dissimilar to what occurred in the industrial development of the Scandinavian countries as their extractive industries declined.

R&D remains very important in the sector, with a number of firms interviewed holding patents either locally or in some cases in the U.S. Patent and Trademark Office and European Patent Office. In fact, one firm has set up an R&D company that is a separate entity from the main firm, so it would not lose the intellectual property were it ever to sell the main company. Another medium-size firm with 60 employees holds 20 patents in various locations, including South Africa, the United Kingdom, and the United States. It has used its own retained profits to apply for and renew these patents.

Channels of technology transfer. Interviews with eight mining equipment companies showed a clear link between exporting and technological development. In all cases, exporting relied heavily on the development of technology. In most cases, exporting entailed significant technological developments, which were sometimes embodied in patents but frequently were safeguarded in more informal ways. Technology absorption tends to focus largely on improving and enhancing application. On the product side, new products tend to be improvements and adaptations to meet the demands of new situations for example, spiral washers for use in oil sands in Canada rather than in coal deposits in South Africa.

Many of the specialized products produced by these firms were developed locally and are globally competitive, recognized for quality and presently cost competitive with respect to their European competitors. However, technological development clearly underpins and is a necessary component of success in export markets in this sector, with firms investing in R&D for continuous product improvement and new product development. At the same time, competition in export markets is intense in this area and consequently fuels further technological development.

Acquisition of technology from international MNCs. The firms interviewed all had a high technological content in the products they produce. They supply to multinational companies that have high demands for quality and standards. As an example, one of the firms acquired technology from France that enabled it to penetrate new export markets in Europe. Linking with international companies has enabled a large amount of technology transfer.

Strong local links with mining companies and research institutes. This sector differentiates itself from the others by the extent of local expertise that is available in South Africa as a result of past competencies built within the mining industry. Though all the firms interviewed invested in in-house R&D, in some cases smaller firms have collaborated with large mining companies to conduct research when their own in-house capacity fell short. Another firm that licenses welding technology from the Welding Institute in the United Kingdom has also been collaborating with the Nelson Mandela University engineering faculty in the area of aluminum welding.

Constraints to enhancing technological competencies. South Africa's capital goods sector has been built on the competencies of the mining sector, which attracted skilled personnel to the sector and where industry-linked research was a priority. However, these competencies are gradually being eroded because of the inadequate skill supply and the declining industry-research links.

Skill shortages. The skills and competencies available for the mining sector have declined in part because of the migration of skilled workers abroad. Firms in the skill-intensive capital goods sector identified the shortage of skills as a major constraint. Skill shortages exist at the managerial, artisanal, and technical levels such as welders and boilermakers.

This skill shortage is being exacerbated by aggressive recruiting on the part of competitor firms, especially in Australia, which have been very successful at recruiting skilled South Africans. By contrast, South African firms complain that they have major problems in securing the necessary work permits for expatriate labor. This situation affects both locally owned and, particularly, foreign-owned firms operating in South Africa.

The ES data provide further evidence of the constraints faced by capital equipment producers and exporters. The 2003 ES had data for 84 capital goods firms, and the 2007 ES had data for 145 capital goods firms. In the 2003 ES survey, exporters and nonexporters, as well as foreign-owned and domestic firms in the sector, cited skill shortage as the top constraint. In the 2007 ES, firms cited crime as the top constraint, followed by skill shortage.

On the supply side, worker training is poorly provided. Firms regard the SETA system as ineffective and inflexible. Exporters of mining equipment believe that the number and quality of mining engineers and related skills are on the decline. The universities of the Witwatersrand and Pretoria, which were once recognized as premier institutions for producing mining and related high-level skills, have both seen significant declines in their capacities in this regard. By contrast, Australia has seven universities offering mining engineering programs.

Firms have responded to the shortage of skills by recruiting skilled personnel from abroad. Some foreign-owned firms had to bring in more staff from their operations abroad than they would have done if the skills had been available locally. However, obtaining work permits has been cumbersome, and the main response has been for firms to train locally. Training is fairly widespread in the industry, particularly on the part of foreign-owned firms. Data from the 2003 ES show that 76 percent of foreign firms in the sector offered training for an average of 42 days for a skilled worker and 50 days for an unskilled worker.

The skill shortage is leading some companies to move some of their operations overseas. These firms are shifting their more labor-intensive operations outside South Africa, notably to China. For example, one of the largest South African operations now fabricates 20 percent of needed capital equipment in China, whereas previously equipment was produced exclusively in South Africa. A number of firms, particularly the larger operations that can manage extensive outsourcing, predict that sourcing product from outside South Africa will increase rapidly. Of particular concern here are the wage rates of technicians and artisans, who are far more highly paid in South Africa and, moreover, are in very short supply.

But manufacturing operations are not the only ones that are likely to be relocated outside South Africa, because the same trend is evident with respect to design and development work. One of the largest South African operations undertaking extensive design and development has decided to develop a new center abroad and predicts that in several areas it will have its main design center in Australia. Some of the major mining companies are similarly locating their R&D activities abroad. The country favored is almost invariably Australia, with the main attraction being the availability of highly skilled labor and links to well-funded research centers.

Declining industry-research links. Only a few of the firms that were interviewed engaged with universities for their research. Apart from accessing the CSIR for testing and verification, industry engagement with the science councils is limited. Moreover, the few links that do exist are said to be becoming more limited over time. This scarcity is true particularly in relation to mining, but also with regard to metallurgy and metal refining.

A widespread view is that although both the CSIR and Mintek (South Africa's mining research organization) have some capacity, they have experienced a clear deterioration over time. Skilled personnel have been lost and a number of programs closed, particularly at the CSIR; the CSIR is said to have hardly any research capacity remaining in relation to mining. The deterioration in publicly funded research for mining, metallurgy, and related activities in South Africa has resulted in firms making much more use of privately funded research. Significant growth is seen in local research consultancies that undertake research or provide specialist consultancy services to the industry. In addition, local firms are increasingly accessing publicly funded research institutions and universities located abroad, particularly in Australia.

Access to finance. A number of the firms interviewed complained that export finance was limited, which sometimes inhibits incentives to increase their technology-related investments as well as their access to the global knowledge pool. The export council SACEEC, in particular, stressed this concern, and indeed, export finance, particularly preshipment finance, is one of the council's central requests to the government. The council, which works with the government on an ongoing basis to ensure that generic policies and priorities are aligned with the sector development strategy, also regards this lack of export finance as particularly injurious to new exporters that are unable to accept large export contracts because they lack adequate finance. Finance is often not available, and where it is, it is said to be very costly.

Most of the firms interviewed had not accessed government support programs for R&D or innovation-related activities. One firm that did use the government's Support Programme for Industrial Innovation (SPII) in the past found it onerous to access, hence inhibiting the firm from using it again.

New entrants are also constrained by the lack of venture capital available to finance technology-based start-ups. Potential new entrants cannot find the financing they need for commercializing their research or seed funds to launch new products. Very few firms have been able to get preferential finance from the Industrial Development Corporation.

Summary
The capital goods sector is undoubtedly the most technologically advanced sector in South Africa as a result of the strong research and technology base that was built before 1994. The local value-added content is as high

as 90 percent in this sector, which reflects the local competencies in the sector. In the past, important links with Mintek and universities led to close collaboration with industry to develop technology. A decline is now perceived in the local competencies available, however, because of the large number of skilled professionals moving abroad (especially to Australia) from the sector, which has in turn led to a decline in the research competencies in the local universities. These declines could imply that South Africa's capital goods sector is possibly losing its global competitiveness to countries such as Australia that continue to increase their competencies in the sector.

South Africa: Chemical Sector

Globally, the chemical industry is characterized by three distinct features: it is technology intensive, capital intensive, and reliant to a large extent on high-quality inputs (because of its sophisticated value-adding chain). The South African chemical industry was initially developed to provide a source of synthetic fuel for mining, later branching into chemical feedstocks and intermediates. Although this specialized industry proved particularly important before 1994, the isolationism and protected environment enabled production at low productivity levels. Isolated from international competition and burdened with high raw material prices because of the cost of producing oil from coal and associated substantial import tariffs, locally processed goods were not competitive in the export market. With the end of isolationism, the industry has been forced to rapidly restructure and improve competitiveness.

Currently the South African chemical sector is the largest in Africa and contributes about 5 percent to South Africa's GDP and 25 percent to its manufacturing sales.[8] The sector is reasonably diverse and is generally divided into four broad categories: base chemicals (including the petrochemical building blocks and inorganic chemicals); intermediate chemicals (ammonia, waxes, solvents, phenols, tars, plastics, and rubbers); chemical end products (processable plastics, paints, explosives, and fertilizers); and specialty chemical end products (pharmaceuticals, agrochemicals, biochemicals, and additives).

The industry is a substantial contributor to South Africa's manufactured exports, comprising 17 percent of total manufactured exports (table 2A.7 in annex 2A). The top export destinations in 2009, shown in table 2A.8 in annex 2A, were Zimbabwe (15 percent), the United Kingdom (13 percent), and Mozambique (10 percent). However, the large majority of those exports are basic and low value added. Moreover, the overall trade account for the sector has a significant negative balance, with the exception of a surplus in a few areas, notably polypropylene. Furthermore, interviews suggested that many chemical exports are currently unprofitable, with some firms exporting at a loss to offload surplus stock produced to achieve scale economies. South Africa's chemical industry is relatively small by global standards. The industry sharply differentiates South Africa from major emerging market competitors in Brazil, China, India, Indonesia, and the Persian Gulf. Because of the increased infrastructure requirements in Brazil, China, and India, the chemical industry in these three emerging economies is increasing three times faster than in the United States. In India, the chemical sector is underpinned by the plastics and fertilizer industries and in Brazil by the agrochemical industry.

Technology Absorption in the Chemical Sector

The industry is dominated by the basic chemicals subsector, whose liquid fuels, olefins, organic solvents, and industrial mineral derivatives together account for about 31 percent of chemical production in the country and make up most of the exports in this sector. They are low-value-added and high-volume products. Interviews suggested a general perception in the industry of little payoff for exporting firms to upgrade and enhance their technological capacities.

Sasol, the largest company in the chemical industry and indeed South Africa's largest industrial company, undertakes extensive R&D and is itself responsible for a significant share of South Africa's total business sector R&D. Sasol is the largest employer of engineers in the country and has global leading-edge technology in a number of areas. However, Sasol is an exception among chemical industry companies. In the ES, R&D was very limited for firms in the chemical industry, with a maximum value of 3 percent of sales and an average R&D figure for the industry as a whole of only 0.3 percent.

Petrochemical production in South Africa is largely centered on the Sasol II and Sasol III plants at Secunda and the majority-Sasol-owned Natref refinery at Sasolburg, where Sasol generates various feedstocks and olefins. Using the Fischer-Tropsch process to convert coal into oil, Sasol produces about 2 million tons annually of a range of olefins for the petrochemical industry. About 0.6 million tons of olefins are used by the chemical industry, and the remaining 1.4 million tons are used in fuel production.[9] When compared with international petrochemical plants based on natural gas or ethane, the local synfuel plants tend to be less competitive. Consequently, Sasol is now importing natural gas piped through to the Secunda site from Mozambique.

Tertiary chemicals include processable plastics, paints, explosives, and fertilizers, which make up only 5 percent of the country's chemical production. Subsumed under this category are specialty chemical end products, which tend to be lower volume, and higher-value-added products. One of the firms interviewed produces such specialty chemicals and fertilizers. It focuses on the upper end of the market, not competing in the commodity end of the business. It employs the largest group of agronomists in southern Africa, and possibly in all of Africa, and has a very good understanding of soil dynamics through its dedicated laboratories. The company makes extensive use of diagnostic technologies, global positioning systems, and the like. Similarly, in mining it focuses on the upper end of the market, that is, optimizing value from resources rather than supplying the cheapest technologies. Investment in those technologies amounts to about 3 percent to 4 percent of turnover (80 agronomists; 50 engineers). The company also has a few worldwide patents (electronic detonators and fertilizers, in particular calcium nitrate).

Channels of technology transfer. Local innovation in this sector is largely focused on marketing and slight modifications of products and processes. Links with international firms as well as vertical links with top South African firms are important channels of technology transfer.

Licensing of production technology from international MNCs. Production technologies are usually imported by means of technology licenses from a top technology vendor, which is typically one of the established multinationals. One of the firms interviewed imported state-of-the-art technology from a leading specialist supplier of chemical plant and equipment in Japan.

Domestic links with top South African firms in the industry. Many of the global players in the field (Dow, DuPont, and Asahi) have vertically integrated operations, which over the years have commercialized their own in-house technologies. Levels of skill and experience are vital in such an operation, though not insurmountable if a technology provider is committed to supporting the venture. One of the firms interviewed had succeeded in establishing domestic links with top South African companies, which in turn had their employees trained by Japanese technicians in key operations.

Constraints on enhancing technological competencies. The skill gap, especially in terms of a shortage of skilled chemical engineers and inadequate feedstock, further constrains a movement toward more value-added products in this sector.

Skill shortages. To process more inorganic chemicals, South Africa will require more knowledge-intensive (for example, engineers, artisans) and service-intensive (for example, accountants) skills. Highly skilled chemical engineers are scarce in South Africa, rendering substantial progress in this area infeasible in the short term. The shortage is also highlighted in the ES data, where 40 percent of the firms surveyed in 2003 and 22 percent of the firms surveyed in 2007 cited labor skills as a major constraint.

Insufficient industry-research links. Few firms appear to have strong relations with the universities. However, one of the firms interviewed had invested heavily in its own technology to produce polyethylene from bioethanol. This firm had its own R&D laboratories that oriented their research toward applications while sourcing fundamental research from universities. The firm also has dealt with science councils and institutions over the years, but this relationship was not robust, especially given the science councils' and the government's insistence on retaining intellectual property rights. This restriction on intellectual property was said to act as a major deterrent to collaborative research.

Lack of competitive feedstocks. Exports in this sector embody very little technological content. A general perception in this sector is that upgrading has little payoff. Feedstocks are high priced relative to the oil-based feedstocks used by the industry's international competitors. South Africa also lacks access to competitively priced raw materials (oil, wet gases, ethane, chlorine, and aromatics). In the organic sector, as long as oil is not available domestically, little rationale exists for engaging in large-scale innovation processes geared to export markets because the feedstocks are not competitively priced. In the inorganic sector, where products are derived substantially from mining, South Africa seems to fare pretty well, given its natural resource endowment and associated corporate capabilities. According to one respondent, however, the mining industry is unlikely to invest in more upstream process innovation because the industry is focused on maximizing raw material output for export. Moreover, minimum efficient scale of production is high, and given the limited domestic and regional markets, combined with South Africa's distance from potential export markets, entrepreneurs are unlikely to find the capital to make that investment.

Future prospects. Given the centrality of the cost of feedstocks to securing a competitive position, South Africa's best prospects are in areas in which it has a feedstock advantage. Three areas are significant in this respect.

The first is to focus on inorganic chemicals using mineral feedstocks. Mineral feedstock options include gold, manganese, chromium, vanadium, copper, antimony, phosphate rock, uranium, titanium, and fluorspar. Fluorspar appears to have particular potential because South Africa has large reserves and exports considerable fluorspar (Government of South Africa 2010a, 62).

South Africa also is a major exporter of polypropylene. In the second area, one of the key projects of the Industrial Policy Action Plan (IPAP) is to use polypropylene for downstream products in areas such as injection molding—the polypropylene conversion project. The IPAP estimates that this project will cost about R 1 billion (approximately US$147 million) and create up to 22,745 new jobs (Government of South Africa 2010a, 62).

The third area is in biomass. Sugarcane grown in South Africa contains more sugar than that grown in Brazil. Brazil currently leads the world in producing polyethylene from bioethanol. It has been suggested that Sasol's technology for gasification, synthesis, and coal separation can be readily applied to biomass feedstocks, either on a stand-alone basis or in combination with biological processing. In fact, the Industrial Development Corporation and the Energy Corporation have already invested R 3.2 billion (approximately US$457 million) in two biofuel projects that collectively should produce about 190 million liters of bioethanol from sugarcane and sugar beets (Van Zyl and Prior 2009).

Enhancing South Africa's production of chemicals using any of these three feedstocks will entail considerable capital investment. It will also entail considerable technological challenges. Interviews suggested that securing the requisite capital and obtaining the necessary skills to enhance technological capacities will be the major challenges.

Summary

Given the unavailability of oil domestically, South Africa's competitive advantage lies in inorganic feedstocks in the chemical sector, which are low-value-added and high-volume products. Technology is accessed through licensing from multinationals in the sector. However, the sector faces skill shortages, with a scarcity of chemical engineers in South Africa and very limited industry-research links. Furthermore, the lack of competitive feedstocks stifles the incentives for innovative activity in the sector.

South Africa: Software and Information and Communication Technology Sector

South Africa's information and communication technology (ICT) market is the largest in the Middle East and Africa (MAIT 2009). Internet penetration rates in South Africa average about 8.6 users per 100 people, and personal computer ownership is about 8.5 computers per 100 people.[10] The number of secure Internet servers per million people is 40.4, whereas the average for Sub-Saharan Africa is only 3.1. In 2008, South African ICT expenditures accounted for 10.1 percent of GDP (World Bank 2010c). The sector itself employs upward of 300,000 people, with US$8.7 billion spent on software in 2007, US$1.8 billion on hardware, and US$3.6 billion on services (Government of South Africa 2010b).

Although most major international vendors have a presence in South Africa, local firms supply most of the country's ICT needs. The South African government is the largest purchaser of ICT products, accounting for 15 percent of revenues (MAIT 2009). Moreover, many of these firms have long provided software and services to other African countries, capitalizing on their knowledge of local environments and their lower cost.

Other African markets account for a great deal of ICT service exports. By value, however, Europe is the main trading partner and the United Kingdom is the main destination. As can be seen from table 2.1, exports of ICT goods constitute 1.6 percent of total goods exports, and ICT service exports account for 3.2 percent of total service exports.

Technology Absorption among Firms in the ICT Sector

South Africa has a competitive advantage in the area of embedded software design and development of custom turnkey solutions. It is known

Table 2.1   South African ICT Exports

Category	2000	2008	Sub-Saharan average
ICT goods exports (percent of total goods exports)	2.0	1.6	0.9
ICT service exports (percent of total service exports)	2.9	3.2	n.a.

Source: World Bank 2010c.
Note: n.a. = not available.

for developing niche applications for specific vertical industries, such as mining, finance, government, and communications. With respect to structure, estimates show that packaged applications account for 41.8 percent of the software market, application development and deployment are 24.8 percent, and systems infrastructure software is 33.4 percent (Government of South Africa 2010b).

The sector has recently accelerated the adoption of international best practice models, such as Capability Maturity Model Integration (an advanced process improvement approach), and embarked on a number of cross-country collaborations. For instance, Wits University's Joburg Centre for Software Engineering has entered into a formal partnership with the Software Engineering Institute in the United States to drive process improvement initiatives in the South African ICT sector. In the sector itself, approximately 50 percent of employees are computer professionals with formal tertiary qualifications in ICT. Electronics and hardware technicians and engineers make up another 30 percent. The sector is clearly highly skilled and specialized. The success of the South African ICT sector is therefore directly linked to its innovative capacity, especially in the areas of systems integration and customized software.

For this case study, interviews were conducted with two relatively large and one medium software firms. One of these firms provides specialist enterprise software solutions for mobile financial services, one provides financial service software, and one manages process control software for major installations.[11]

Most of the mobile solution provider's exports are destined for the rapidly growing mobile market in developing countries. Its market includes 22 countries in Africa and the Middle East, 5 in Asia and the Pacific, and 2 in Latin America. The financial service provider operates in 51 countries and is the world's leading service provider, with a deliberate focus on both developed and emerging markets. The process control software provider currently has numerous clients in the United States, and to a lesser extent in Angola, Australia, France, Japan, Saudi

Arabia, South Africa, Trinidad, and in South America (particularly in the oil refinery sector).

Channels of technology transfer. Continuous investment in R&D is a key element to remain globally competitive in the ICT sector. In addition to investment in R&D, firms have leveraged on their networks abroad, including their client base.

Investments in R&D. In terms of technology, the firms have all adopted fairly aggressive strategies. The mobile service firm has a portfolio of 10 patents, and despite being an industry leader, it continues to research ways to improve its portfolio. This effort includes increasing its R&D budget from 15 percent to 35 percent of revenue and tripling the number of developers. Because the software industry is characterized by continual releases of new versions, the mobile service firm also invests heavily in management to interface with clients. The financial service provider focuses more on continuous improvement of base products than on new offerings and has more than 100 developers in the Cape Town office dedicated to this task. It has no patents but relies on copyright licenses and client license agreements to protect its software assets. Likewise, the technology of the process control software provider is not patented and is protected only by normal software copyright.

The large mobile service firm was ineligible for government support through the SPII initiative but had availed itself of the R&D tax allowances. One of the other firms has accessed the SPII but commented on finding the procedure administratively burdensome.

Links with international clients. With respect to technology and exporting, all the firms feel that exposure to a wider field of demanding clients gives them a competitive edge. The perception is that having a multicountry client base increases a firm's credibility in gaining new international clients. However, only the mobile service provider considers exporting to be a major factor in technology upgrading. The key to all the firms' success has been the ability to attract high-quality, skilled software developers both locally and from abroad. Incentives to invest in technological innovations are sometimes spurred by client demands, as was cited by one of the firms. A retailer in the United Kingdom led the financial service firm to develop a major gift-card system, which required the firm to develop the requisite technology. That technology has subsequently taken off with many other clients around the world.

Constraints on enhancing technological competencies. The ability to attract and retain skilled ICT professionals is central to the sector's firms so they can keep up with global developments in the sector. The skill shortages in South Africa and the consequently rising wages thus pose a significant constraint to this sector. In addition, high broadband costs constrain the sector's firms.

Skill shortages. As highlighted previously, the competitive edge of these firms has been their ability to attract highly skilled software developers. Initially South Africa also had a cost competitiveness that is now being eroded as a result of its skill shortages and increasing wages. The skill shortage was highlighted by all the firms interviewed. Generally this related more to quantity than quality, though one respondent noted a marked decline in the quality of graduates from second-tier institutions. Also, so few first-tier applicants are available that some firms are forced to employ workers from India at a substantial relocation premium. Moreover, labor turnover is significant, with one of the firms citing turnover rates of 10 percent per year.

The skill shortage is compounded by black economic empowerment (BEE) policies, which come into play when servicing South African clients.[12] The skill shortage is even more pronounced in previously disadvantaged groups, making qualified black ICT professionals difficult and costly to hire. One firm was very concerned about the possibility of BEE being extended to foreign client service, a move that could force disinvestment in South Africa. The lack of regulatory certainty in this regard is already perceived as constraining development, suggesting that the government should look into at least temporarily relaxing BEE stipulations in underrepresented sectors. Two of the three firms have operations in India, where they had recently hired most of the software developers because they were unable to successfully fill vacancies in their local South African offices.

Intellectual property ownership. One firm noted that the restrictions on the movement of intellectual property, specifically the lack of freedom to move it offshore, act as a significant disincentive to technology upgrading. Permission has to be obtained from the South African Reserve Bank and is rarely granted. As a result, a number of South African ICT firms with potential global prospects have not been able to move their intellectual property offshore, which makes accessing international venture capital to fund expansions abroad very difficult, if not impossible.

High broadband costs. The major infrastructural constraint relates to bandwidth, which is both costly and of poor quality. The government should consider policies to improve the quality and reduce the price of telecommunications. The new state-owned telecommunication infrastructure provider, Broadband Infraco, could help increase available bandwidth capacity and lower cost. Broadband Infraco has taken over 12,000 kilometers of the long-distance fiber-optic network once owned by Transnet and Eskom. Broadband Infraco is also part of the consortium building the West Africa Cable System (Mashego 2010). The benefits of these initiatives have yet to be realized, however.

Summary

The software and ICT sector in South Africa has been built on the competitive edge of being able to produce innovative solutions, the key to which has been the ability to attract highly skilled software developers. This competitive edge is now under severe threat, with a decline in the number of ICT graduates available in the local market, implying that many of these firms have to employ expatriate professionals, mainly from India at a relocation premium. The high broadband cost, at US$26 per month (compared with US$6 per month in India and US$19 per month in China), puts further pressure on these firms by straining them financially and inhibiting the technology development process. The restrictions on the movement of intellectual property, specifically the lack of freedom to move the intellectual property offshore, act as another significant disincentive to technology upgrading among these firms.

South Africa: National Innovation Survey and Enterprise Survey

The findings of the empirical analysis conducted using the National Innovation Survey (NIS) and the ES for South Africa (detailed regression results are shown in tables 2C.1–2C.11 in annex 2C) corroborate the findings of the industry-level case studies. As described in chapter 1, technology absorption depends on the incentives for firms to invest in technology upgrades, their access to the global technology pool, and the absorptive capacity of the firms.

Incentives for Firms to Absorb Technology

The NIS data analysis shows that intellectual property rights (IPR) positively and significantly affect the introduction of new product in both the full sample and the subsample. Regarding effects on new process, the impact of IPR is seen only in the innovator subsample; no impact is seen on changes in organization, and surprisingly, a negative impact is seen on marketing. This negative impact is evidenced by the constraint that some ICT firms in the case study analysis voiced in terms of the existing IPR regime in South Africa. An inflexible IPR regime has stymied the extent of technology absorption. As a result, a number of South African ICT firms with potential global prospects have not been able to move their intellectual property offshore, which hinders or even prevents them from accessing international venture capital to fund expansions abroad.

Collaboration with other firms in the enterprise group—that is, suppliers of equipment, materials, and clients or customers—increases the propensity of a firm to introduce a new product, a new process, and organizational changes in the innovators sample (unavailable in the full sample) but has no effect on marketing. The importance of successful industry-research collaboration is further reflected by firms interviewed for the case studies, that cited the lack of adequate industry-research collaboration as an important constraint to improving the level of technology absorption.

The ES analysis for the 2007 sample shows that firms that experience competition from informal firms also have a higher probability of introducing new products, which is indicative of their need to differentiate themselves in the market (table 2C.11 in annex 2C).

Financial constraints positively and significantly affect the introduction of a new process in the full sample but not in the subsample; they have a strong effect on organizational changes, both in the full sample and in the subsample, but no effect on marketing. Financial constraints could be important in determining the firm's incentives to adopt new technology as well as in determining the firm's absorptive capacity. The regression suggests that when a firm experiences internal or external financial constraints, it has a higher probability of introducing a new process, possibly an indication of the cost-alleviating new processes being introduced to achieve greater firm efficiency. Investments in technology upgrades and R&D require a greater level of financial flexibility than firms can afford. For that reason, effective public support instruments play a vital role in increasing the levels of technology absorption and innovation.

Different types of finance are generally used for research, innovation, and technology development, depending on the market readiness of the technology and the growth stage of the company. In most countries, public funding and, to a lesser extent, funding from large corporations are used to finance basic or long-term research with no immediate market applications and high positive externalities. The ES data show that sample firms in South Africa that have greater access to credit have a positive and significant relation to introducing new processes.

Access to the Global Technology Pool

Exports have a positive effect on the probability of absorbing technology: the result is significant for all four dimensions of technology absorption in the innovators subsample and for only changes in marketing in the full sample. This finding indeed is consistent with those of the country case interviews, in which exporters tend to invest in technology upgrades and R&D to become even more competitive in the global market they compete in. Although in some instances exporters have the impetus to invest more in technology absorption, many firms, because of their technical advancements, can venture into exporting into new markets.

In addition, subsidiaries of foreign firms have a higher probability of introducing organizational changes, but surprisingly, the surveys show no significant effects on any of the other variables' indicators. In the ES analyses for both 2003 and 2007, when firms engage in a joint venture with a foreign firm, a large positive and significant relationship exists with the probability of introducing new products, suggesting that cultivating interfirm relationships is important. Joint ventures have often been credited with being an effective form of technology transfer through FDI, because they involve a local partner.

Absorptive Capacity of Firms

In the full NIS sample and the innovators subsamples, medium and large firms have a higher probability of introducing a new product or process or changes in marketing or organization, compared with small firms. In the ES sample, large firms demonstrate a large and statistically significant positive impact on the probability of introducing a new product compared with small firms.

The education variable in the NIS (the share of tertiary degree holders in the firm) is not significantly correlated with any of the technology absorption indicators. The NIS did not contain data on the education and experience of the managers, a variable that is significant in the ES analysis. Questions about the education of the manager and other variables that would allow more insight into entrepreneurial skills were not asked.

In the 2003 ES data (table 2C.10 in annex 2C), the manager's education and years of experience are both positively and significantly related to the propensity to introduce a new product or a new process, providing evidence of the importance of skilled labor to undertake innovative activity. Human capital has an obvious bearing on the production of new knowledge in business practices. Human capabilities—some of them acquired in the formal process of human capital formation and others affected by that mode of training—have a significant bearing on the frequency and ubiquity of innovative economic behavior and on the supply of entrepreneurship.

Whereas managerial skill levels have a positive and significant relationship to the probability of firms introducing both new products and new processes, worker skills and training levels do not have the same relationship. This difference could in part reflect the low effectiveness of training programs in South Africa. Furthermore, despite the significant resources devoted to education, inequalities in terms of learning outcomes and educational attainment persist. Compared with other countries, the overall quality of the South African educational system is poor, hence potentially harming firm competitiveness.

One of the most important factors affecting the introduction of both new products and new processes is R&D (for 2003 ES data). The magnitude and significance of R&D, as reflected in table 2C.10 in annex 2C, demonstrate that R&D is likely the most important contributing factor to both product and process introductions and therefore ought to be given considerable weight in any discussion of policy implications associated with this analysis. As stated earlier, the role of R&D is not confined to innovation but is crucial for technology absorption, adoption, and imitation as well. Griffith, Redding, and Van Reenen (2004) show that R&D enhances technology transfer by improving the ability of firms to learn about advances on the technology frontier. Thus, R&D is important not only in the process of catch-up but also in that of directly stimulating innovation. This conclusion is an important result for South Africa, which exhibits the unique characteristic of R&D expenditure dominated by the private sector—a trend dissimilar to that in other developing countries.

In the NIS sample, R&D, which is only available for the innovators subsample, also has a positive and significant impact on the introduction of a new process, a new product, and organizational changes, but not of marketing changes. The case studies showed a similar trend: firms that invested heavily in R&D were able to produce new products that were subsequently patented. Those firms had a separate R&D budget allocation.

R&D is an input both for the production of innovation and for technology absorption. Yet before proxies for innovation and technology absorption became widely available, R&D served as a proxy for innovation. Patents have been used for a long time as measures of innovation, but they too did not become easily available until the U.S. Patent and Trademark Office was computerized around the 1980s.

Using the NIS data, one can explore the determinants of firm-level R&D. The variables that show significant correlation with R&D are exports and a variety of IPR variables (table 2C.8 in annex 2C).[13] Interestingly, patents secured in South Africa are positively correlated with R&D, but patents held outside South Africa are not. This distinction may be because the mining and related products industry has had its own local patents for many years, but it also may be an indication of a certain lack of openness of the economy, because foreign patents are not important for local R&D. The reliability of this finding, however, is doubtful: only 88 companies (2 percent) had ever secured a patent in South Africa, and 63 (2 percent) applied for a patent outside South Africa. Foreign subsidiaries have a lower probability of engaging in R&D compared with domestic subsidiaries, indicative of the fact that R&D is conducted at the parent company, whose headquarters are elsewhere, whereas domestic subsidiaries invest locally in R&D.

Collaboration with other sister firms in an enterprise group, with suppliers of equipment and materials, and with clients or customers is positively and significantly correlated with R&D. Managers were asked which type of cooperation partner was most valuable for their firm's innovation activities. Other firms in the enterprise group, suppliers of equipment, and clients were frequently mentioned by firms, and universities or technikons and government or public research institutes were mentioned less frequently (mentioned by only 6 and 11 firms, respectively).

The NIS allows the comparison between new-to-the-market and new-to-the-firm innovations. As Mohnen, Mairesse, and Dagenais (2006) suggest, new-to-the-firm innovations are more likely to simply be imitations of existing technology, whereas new-to-the-market innovations are more likely to be true innovation (rather than imitation). New-to-the-market activities are positively and significantly correlated with R&D and the acquisition of external knowledge (table 2C.7 in annex 2C). Although exports lower the probability of new-to-the-market innovations, foreign subsidiaries conduct more new-to the-market innovation. Although the share of employees with tertiary education was not correlated with any of the technology absorption variables discussed above, it is positive and significantly correlated with new-to-the-market innovation.

Managers' Perceptions of the Constraints to Technology Absorption

The factors hampering activities to further technology absorption are summarized in table 2C.9 in annex 2C. The Likert scale[14] indicates the relative importance of the variable, ranging from not relevant to low, medium, and high levels of importance.

Of the constraints listed, the greatest number of firms that attributed a high degree of importance to the reason for not innovating were those that perceived the market to be dominated by established firms (26 percent).[15] It is therefore unfortunate that a variable for the degree of competition within each sector was not included in the NIS.

The next highest constraint was a lack of internal funds (25 percent), followed by innovation costs being too high (20 percent). Unfortunately, only 5 percent of the firms received public support for innovative activity. Most studies conclude that government R&D support leads to additional private R&D, innovation expenditures, or innovation outputs and not to crowding-out of private R&D by public R&D support.[16] The forthcoming study (World Bank 2011a) presents concrete suggestions to increase the volume of matching grants to enterprises and other instruments to support commercialization of research.

Lack of qualified personnel came in as the next highest constraint at 17 percent. This result is somewhat surprising, because one would have expected skill shortage to be the most important constraint and certainly to be mentioned by more than 17 percent of the firms. Yet managers assessing the financial constraint compared to the skill shortage constraint may see funding as a binding constraint in the immediate term and lack of skills as an obstacle to longer-term development.

These constraints echo the findings from the case study analyses and suggest the urgent need for greater public policy attention to alleviate these obstacles for the private sector in South Africa if the country is to foster greater technology absorption and thereby enable productivity and efficiency gains in the future.

Namibia: Agroprocessing Sector

The exporters interviewed in Namibia were all in the agroprocessing sector, given that the sector constitutes more than half of Namibia's exports.

Agroprocessing is technically a manufacturing activity, although many processing firms often produce and harvest agricultural raw materials themselves, often blurring the boundary between agriculture and manufacturing. In Namibia, the agroprocessing sector is particularly important, accounting for a little over 70 percent of the total manufacturing output, with fish processing averaging 20 percent, food and beverages 45 percent, and meat processing 6 percent in the period 1995–2006 (table 2A.9 in annex 2A). Fishing represents Namibia's second largest foreign currency export earner (after mining), and 90 percent of the national fisheries output is exported.

The agroprocessing sector also employs the second largest proportion of the labor force, accounting for 42 percent of the manufactured employment after agriculture, which is by far the largest employer. In recent years, however, both fish and meat processing have been declining in percentage contribution as textiles, diamond, zinc, and copper processing have increased. This trend is also evident in the export percentage of these categories in the period 1995–2006. In spite of the decline, the agroprocessing exports still constitute more than half of Namibia's exports.

To better determine the constraints and profile of the industry, this case study used interviews of four firms—two large fisheries, a meat processor, and a brewery. The brewery and the meat processor have no local competitors. Thus, they effectively comprise the entirety of their respective industries. The fish-processing firms are two of the dominant firms in the onshore fish-processing industry and together are responsible for approximately one-quarter of employment in the fisheries sector. Together the four firms interviewed comprise some 3,800 employees, approximately 25 percent of total employment in the Namibian agroprocessing sector. For all four firms, a very large percentage of their product is exported—significantly more than half with respect to the brewery, 85 percent for the meat processor, and over 90 percent for the two fish processors.

The meat processor's prime markets for exports are Norway, Switzerland, Reunion Island, and now the United States. It also exports a small proportion to Asian markets—to Hong Kong SAR, China, and China. Of its total production, the meat producer exports 35 percent to Europe; however, this volume makes up 65 percent of total value because the high-quality meat sold to Europe earns a premium. The Namibian brewery is publicly listed on the Namibian stock exchange.[17] Ohlthaver and List Group, one of the largest conglomerates in Namibia, wholly owns one of the fisheries and has a majority stake in the brewery. The other fishery interviewed is the only privately owned firm in the Namibian fishing industry; it has partnerships with BEE partners who hold 60 percent of the company. The brewery exports about 55 percent of its production from Namibia, with South Africa being its largest export market, followed by Angola and Botswana and small percentages to another 27 countries. Its market segment is selling premium beer to these markets. Europe was the prime market for premium hake products for both the fisheries interviewed.

Technology Absorption in the Agroprocessing Sector

The impetus for investing in technology in all the firms interviewed was to meet the stringent standards of the European markets, which are their main export markets. Many food-processing activities use sophisticated technology such as quality assurance and food safety management systems, giving rise to significant technological learning opportunities (OECD 2010). The use of sophisticated technologies is particularly prevalent where a product is destined for export markets, especially when exported to the markets of developed countries. In the case of Namibia, given the small size of the domestic market and the limited regional market, the developed-country market is the destination for a very large share of the final product. All the Namibian firms visited accordingly used very sophisticated production technologies.

In meat processing, the imperative to use the latest technology arises principally from the need to meet the demanding standards of export markets in Europe and the United States. Some customers also require compliance with green production and corporate social responsibility. As with meat processing, the principal driver for using the latest technologies in fish processing is the demanding standards of export markets. Health standards are particularly important for food products.

In meat processing, continuous efforts are made to upgrade the technology, in particular the recent investment in radio frequency identification (RFID) technology to trace cattle. In fish processing, both companies used state-of-the-art production technologies. These include electronic software to record and monitor production processes, intelligent portioning equipment, and sophisticated freezer systems. Hazard Analysis Critical Control Points is used as a food safety management system. A new factory currently under construction is being equipped with the most sophisticated individual-portions quick-freezer technology.

In brewing, state-of-the-art technology is used at every stage of the production and the marketing and distribution processes. The brewery uses a particular brewing process that has no preservatives and additives in the process and imports all its raw materials from Europe. The brewing is done according to the Reinheitsgebot standard, a German beer purity standard, which permits only four ingredients in the beverage—water, hops, barley, and yeast—thus preventing the use of any additives. Exports and the associated quality and branding of the product, therefore, are the principal drivers of the use of the latest technologies.

Branding and market intelligence. Sophisticated technology is not, however, confined to production. Marketing technology is also critical. Because the firms all produce consumer products, gaining knowledge and meeting the precise demands of consumers are critical to success. One of the fishing firms, in particular, has an innovation strategy that is highly attuned to the opportunities and needs in its overseas markets. The firm has dedicated marketing partners who play a key role in identifying short-term opportunities for premium-priced products. This strategy is a joint venture with five marketing and distribution companies from around the world. The quality of the market intelligence arising from these partners enables the firm to customize its offerings to the particular preferences of individual customers and to adapt at short notice to any changes in their requirements. This firm's strategy is a sophisticated international marketing approach with a dedicated market channel, a highly defined niche market, and a powerful brand.

The brewery, too, is engaged in extensively gathering marketing intelligence, assessing customer needs, and marketing its products accordingly. For beer, the branding stresses purity, the absence of additives, and the mix of German and African influences.

Both fish processors have developed their own distinctive brands, with one company selling exclusively under its own brand, the trademark for which is registered locally. The fish portion sizes are adjusted exactly to what the customer requires. The fish processors stress the superior quality of Namibian hake, by comparison with Chilean, for example. The meat processor also has its own brands and differentiates itself from the world's larger meat suppliers on the basis of its having free-range cattle that use no hormone supplements. Clearly, the branding these companies use shares common characteristics that stress natural ingredients from a country where the environment remains natural and unspoiled.

Channels of technology transfer. Although these firms have very limited formal R&D, they are all investing in adaptations of existing products.

The fish and meat processors focus on ensuring the efficiencies of the production line. The brewery similarly focuses attention on improving the production processes. Though one of the fishing companies has recently set up a new product development department, only the brewery does any substantive new product development. The brewery has about a dozen employees engaged in new product development and estimates that the cost of new product development is 1 percent to 2 percent of turnover. One of the fisheries interviewed recently set up a new product development department. It also recently invested in purchasing new vessels to reduce production downtime, maintenance costs, and fuel costs. As an additional measure to reduce fuel costs, the firm installed its own fuel-mixing facility, which blends two grades of fuel to produce its own heavy marine diesel oil, which is less costly than the normal diesel fuel being used. This blending installation has reduced its fuel costs by 23 percent so far. In addition, the firm has begun selling this blended fuel at a premium to other fishing companies, which generates additional revenues. These initiatives were motivated by the firm's desire to move into higher-margin products, thus reducing costs. However, the firms have not received any financial support to undertake these innovative activities and have expressed a concern regarding the unavailability of any public financial instruments to create incentives for investment in training or private R&D.

Acquisition of machinery from abroad. In meat and fish processing, the acquisition of technology is arm's-length, and ongoing technical agreements are limited. Technology is acquired from global leaders. The meat processor got its RFID technology from Australia and additional machinery from both France and Australia. A critical component of this technology transfer is the consulting services provided by the technology producers from Australia and France, who spent a few weeks in Namibia to train the local workers on the new machinery and equipment. One fisheries firm bought its weight-measuring equipment from Finland, produced by the leader in the field of fish-processing technology.

Technical exchanges with foreign partners. For the brewery, links with foreign partners (Heineken and Diageo) are critical. The partners have ongoing exchanges of technical information and staff exchange programs. Furthermore, the foreign partners engage in training and also regularly send experts to Namibia. These links are critical to the firm's ability to sustain quality. The brewery, in turn, has assisted one of the fisheries (which is wholly owned by the conglomerate holding company) to use a new product tool, which it obtained through its partnership with Heineken and Diageo, to investigate new product opportunities.

Links with international research institutions. None of the firms had any substantive engagement with research undertaken locally—by government or at the local university. The brewery, however, has strong links with Weihenstephan University in Germany, where the brewers obtain their qualifications and where some undertake postgraduate study. Weihenstephan University also audits the quality of the brewery product, hence playing a vital role in the process of technology absorption in the firm.

Constraints on enhancing technological competencies. The Namibian firms interviewed use very sophisticated processing technologies that they acquire almost entirely from foreign vendors. However, a number of factors constrain the further development of technological capacities. Most critical are the limited supply of the natural inputs and the limited supply of skills.

Skill shortages. Lack of the requisite high-level skills particularly limits the ability of firms to extract maximum efficiency from their existing production processes. It also exacerbates the difficulties that firms anticipate if they are to move into more sophisticated products and the requisite adoption of the appropriate processing technologies, which require more high-level skills. Firms complained that they were not able to obtain the skills that they required locally. This situation was particularly stark for the meat processor that had no foreign partnerships and thus faced higher barriers to access technical training services. Technical, artisanal, and plant maintenance skills are particularly in short supply and are very costly. In addition, companies complained that obtaining work permits to hire people from abroad is extremely difficult. For short-term consultants from abroad, work permits are required and are said to be equally difficult to acquire—even to bring in consultants to do the training that accompanies the purchase of new plant and equipment. Firms said they encounter very long delays in the granting of work permits. Even when short-term work permits are granted, the maximum period is three months. This delay is a major constraint to a small economy that has limited skilled personnel available locally.

Limited supply of natural inputs. A limited supply of natural inputs is the most significant factor constraining further investment in enhancing technology. For example, although the meat processor has RFID traceability technology so that it can trace the batch in production, it does not have the technology that would enable it to trace individual cattle. The investment in this technology is desirable, but it is not economically viable at the current production output. Current production, in turn, is constrained by the limited (and indeed declining) supply of good-quality slaughter stock. Similarly, one of the fishing companies has been considering investing in a plant to enable second- and third-stage value addition. However, the stumbling block is securing sufficient volumes of fish.

Something of a paradox is evident here. The short supply of natural inputs provides an incentive to Namibian firms in meat and fish processing to move up the value chain to gain larger margin. This situation, combined with the predominance of exports to sophisticated and demanding markets, has, in turn, required that firms engage sophisticated production technologies. At the same time, the short supply of raw materials economically constrains the investment made to adopt the new technologies that are required if the firms are to move their product into higher-price market brackets. Any significant further technological upgrading in the fish and meat processors will therefore depend heavily on the expansion in the supply of natural inputs.[18]

Insufficient supporting infrastructure to meet standards. The limited availability of raw material inputs has pushed the firms interviewed into higher-value-added products to increase their revenues. The movement up the value chain has, in turn, required that firms meet ever-more-demanding international quality standards.

Firms meeting more demanding standards not only face challenges; they also face a lack of adequate support from public provisioning infrastructure including the lack of accredited control laboratories and insufficient enforcement of standards on the part of the government. One example is the inadequate veterinary services, improvement of which is necessary to increase exports in some high-value niche markets.

The fish processors expressed particular concern that Namibia had not joined the Marine Stewardship Council (MSC), an internationally recognized labeling regime that is consistent with the World Trade Organization and supported by the Food and Agriculture Organization of the United

Nations, and that, at the time of the interview, the government appeared to be moving away from joining the MSC. Both the fishing companies stressed the importance of MSC certification. It is based on the condition of a country's fish stocks, the impact of the fishery on the marine environment, and the sustainability of fishery management systems. Certification can provide numerous benefits to companies and fisheries, including potential for increased sales to customers that want MSC-certified products, greater ability to attract capital investment and joint ventures, reduced market volatility, lower uncertainty costs, and a more sustainable fishery.

In addition, the interviewed firms faced other constraints, such as logistical bottlenecks, high energy costs, and labor market rigidities, that decreased the investment returns on new technology and processes.

Summary

The firms interviewed in Namibia all had one thing in common: their unique selling point was being able to produce premium products and therefore having to invest in new technology to maintain the high-quality standards demanded by their clients. In addition to catering to their clients, the firms faced low volumes of raw materials, which forced them to move into higher-value-added products to compete effectively in the global market, which was an added incentive to invest in technology upgrading. However, they did face two serious constraints—the low skill base in Namibia, which was compounded by the stringent regulations for obtaining work permits for expatriate labor, and the low volumes of available raw material, making continuous process upgrades economically unviable. The firms that had partnerships with foreign companies were able to tap into those partnerships for training and product development, whereas the meat processor relied on the transfer of machinery and equipment from abroad and the accompanying consultants for its worker training and technology transfer.

Mauritius: Manufacturing Industries

Although manufacturing still dominates exports from Mauritius, contributing to 80 percent of total domestic exports (Government of Mauritius 2010), financial services, offshore global business, and ICT have developed as the economy's new sectors of growth. The largest share of manufactured exports from Mauritius is low-technology-intensive textile products. As of 2008, approximately 40 percent of Mauritius's

US$1.5 billion exports have been attributed to apparel products. Sugar and sugar confectionary constitute the second largest segment of exports, accounting for approximately 11 percent of total exports. The remaining exports are fragmented across a broad range of products, including processed food, and electrical machinery and equipment.

To increase competitiveness in the domestic and export markets, Mauritius's "Industrial and SME Strategic Plan 2010–2013" (Government of Mauritius 2010) focuses on industrial development with an emphasis on entrepreneurial and innovation-led growth. Though Mauritius is recognized as a business-friendly destination and the top-ranked African country (ranked 20th worldwide in *Doing Business 2011* [World Bank 2011c]), the recent loss of preferential trade treatment in textiles intensifies the requirement to diversify the economy away from sugar and textiles. The Mauritian government has been trying to remodel itself into a sophisticated offshore banking and telecommunication economy, given its diminishing access to textile and sugar trade preferences. Trading, shipping, insurance, and funds management are also growing in importance (Habyarimana and others 2005).

As of January 2009, 258 companies were operating in the information technology (IT) and business process outsourcing (BPO) industry, employing more than 10,000 people. BPO services represent 40 percent of the IT industry (the largest share of the industry) and employ nearly 45 percent of the total IT-BPO workforce. Software development represents 21 percent of the IT-BPO industry. Aggregate revenues of the IT-BPO industry (excluding hardware and telecommunications) were estimated to be MUR 10 billion in 2008 (approximately US$330 million).

Case study interviews were conducted with 16 exporting firms in Mauritius.[19] Of the 16 firms interviewed, 14 were manufacturing firms from a range of sectors that included food, capital equipment, jewelry, textiles, and plastics, and two were IT firms (to capture the trends in the new ICT and global offshore businesses into which Mauritius has been diversifying). Five of these firms were foreign owned: the two IT firms (100 percent foreign owned), a garment firm (85 percent), a capital equipment firm (100 percent), and a seafood-processing firm (50 percent). The 2008 ES for Mauritius has a total sample of 384 firms, 177 manufacturing firms and 207 in the service industry. The share of firms introducing new products or upgrading existing product lines was a little more than 50 percent of the firms in the sample, and the share of firms introducing a new process technology was 43 percent.

Technology Absorption in Mauritian Firms

A host of incentives encourage firms to invest in technology upgrades and introduce new products and processes. These incentives include the need for greater product diversification, cost reduction, and compliance with global best practice standards to enter new export markets. Given the small domestic market in Mauritius, exports are often a vent for surplus product, and the investment in new technology to make the product adhere to global standards becomes important.

Channels of technology transfer. Although textiles were the mainstay for the Mauritian economy, the phasing out of the Multifiber Agreement (MFA) and the increasing competition from low-cost textile producers from Bangladesh, China, and Sri Lanka forced Mauritian textile exporters to move to higher-value-added niche markets where their competitive advantage lies in higher-quality garments and an efficient turnaround time. One of the textile firms interviewed is a case in point. This firm is Mauritius's largest textile and garment firm. It started operations in 1986, with 20 employees doing basic stitching, and has now grown to employ more than 10,000 people running a state-of-the-art, fully integrated textile plant focused on producing jersey wear. With the phasing out of the MFA and the onset of the African Growth and Opportunity Act (AGOA), the firm was faced with making major new investments in spinning mills to secure duty-free entry to the U.S. market. This history leads to the unique selling point of this firm, which makes its own yarn (using cotton imported from India), and enables the firm to deliver the product within 2 to 4 weeks of the date of order, rather than the 12 weeks that its competitors require, both locally and in the East. The firm's clients are in the United States and Europe (mainly France, Germany, and the United Kingdom) and are willing to pay a premium for the higher quality and the quicker response time.

The firm financed its investments in the machinery and technology required largely through retained earnings, plowing profits back into investment. The firm did take advantage of the government cost-sharing Technology Diffusion Scheme in 1994 to buy some of the equipment required for the fully automated spinning mills. The firm employs 3,500 workers from Bangladesh, China, and India, and a critical ingredient to its success has been the ease with which it has been able to import expatriate skilled labor.

In addition to facing competitive pressure while entering export markets, firms sometimes were forced to invest in new technology as a result of environmental and quality standards. An example is the case of a food-processing firm that had to invest in pollution-free technology because of pressure from environmentalists. Other more common factors that led firms to adopt new technologies were requirements to meet international quality standard requirements.

An example of a firm diversifying into other sectors is a previous sugar-manufacturing firm. The firm has diversified into providing management services to sugar producers in other African countries, which involves continuously investing in new machinery and equipment. That firm is now engaged in the promotion and realization of engineering projects in sugar-related equipment manufacturing. The projects include manufacturing cane diffusers, heavy-duty cane shredders, water treatment plants, and pumping stations. The firm's client base includes Burundi, Cameroon, Kenya, Malawi, Reunion Island, Tanzania, and Zambia and, more recently, the Philippines and other Southeast Asian countries. Leveraging its expertise in the sugar industry, the firm faces no competition in its markets in Africa. The firm's main equipment supplier, which is in France, trains the firm's employees on the new machines.

Acquiring machinery from abroad. The 2008 Mauritius ES asked firms to specify the most important ways that they acquired new technology. As their most important method, the foreign firm responses break down as follows: embodied in new machinery, 50 percent; developed in house, 11 percent; obtained through a foreign license, 11 percent; and developed with suppliers and obtained through hires, 6 percent each. The domestic firm responses were as follows: embodied in new machinery, 46 percent; developed in house, 12 percent; obtained through the hire of key personnel, 14 percent; and developed with a supplier, 11 percent. A graphic comparison is provided in figure 2.3. The dominance of new machinery is obvious for both groups, as is in-house development, development with a key supplier, and the hiring of key staff. As for South Africa, this finding suggests that downstream transfer and turnover are important conduits in Mauritius. With respect to where firms sourced information about new technology, business associations and consultants were important for foreign firms, and business associations and trade fairs were important for domestic firms.

The case interviews suggest a similar trend, with most firms sourcing machinery and equipment from abroad. Just to name a few, the large textile and garment firm bought machinery and equipment from Switzerland, the furniture producer bought equipment from South Africa, and one of

Figure 2.3   Preferred Method of Technology Acquisition, Mauritius

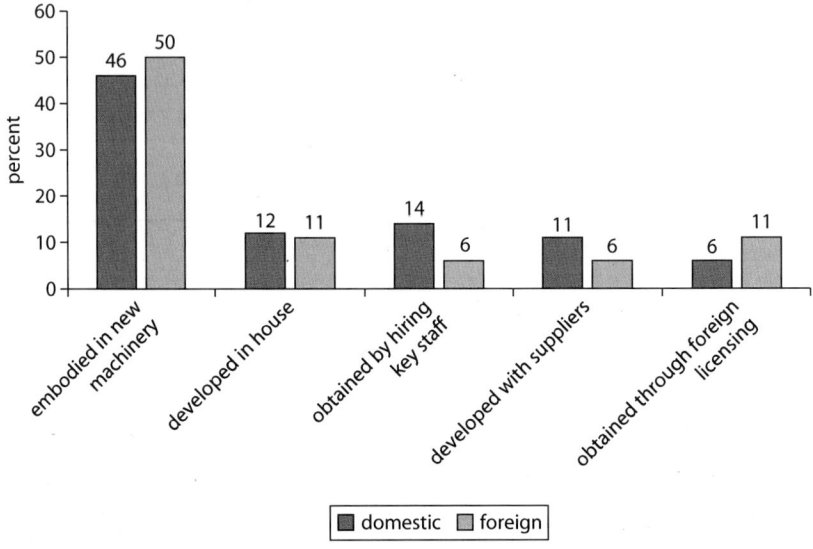

Source: Enterprise Survey (database), 2008 data.

the garment firms bought its computer-aided design equipment from Germany. The seafood processor bought its new machinery and equipment from Germany and Japan. All the firms that bought new equipment received training from the equipment suppliers. Another highly prevalent trait is the purchase of secondhand equipment by smaller firms, caused by the lack of financial resources to buy new equipment.

All the firms interviewed conduct in-house training and use the Industrial and Vocational Training Board for basic training of their employees. They all contributed to the training levy and were satisfied with the reimbursement of this levy upon proof of training provided.

Hiring skilled personnel from abroad. Moving into more quality demanding high-value-added products and investing in new technology require firms to possess skilled personnel. Most critical high-level skills, including dye masters, designers, market specialists, and the like in the textile and garment firms, were imported from India. At the lower end, these firms face the difficulty of securing motivated local labor, particularly to undertake shift work to ensure that production is 24/7 (in the case of the big textile and garment firm). A significant proportion (35–40 percent) of the labor force in the large textile and garment firm is acquired outside Mauritius, mainly from Bangladesh and India. The other garment firms interviewed also employ expatriate labor from Sri Lanka, with one of the firms hiring 1,000 expatriate workers for its total workforce of 2,500 employees. The seafood-processing firm also hired a large number of expatriate laborers from India to work on the new equipment it had purchased.

Both the IT firms interviewed had most difficulty in recruiting tele-marketing agents, IT consultants, business transformers, and IT technolo-gists. They are recruiting expatriates because of the lack of available locals who are trained in the related field. The ease of recruiting foreign ICT professionals was cited as a critical success factor for firms doing business in Mauritius. The flip side, however, is that the shortage of skilled IT pro-fessionals puts upward pressure on wages, making the sector globally uncompetitive.

Learning from exporting. A wide range of technological competencies is represented among the firms interviewed. The level of sophistication was to a large extent determined by the export destination. One could also argue that the level of technological competencies determined the export destinations. Some firms had very traditional technologies and products and exported them mainly to the regional markets, such as the Southern African Development Community (SADC), and Reunion Island. Many of these firms also bought secondhand equipment because they did not have the financial flexibility to buy new equipment, and they were able to make do with that secondhand equipment, which served their needs. In contrast, firms that exported to technologically more advanced desti-nations, such as the United States and the European Union, invested in new technology and equipment to be competitive in those markets and to adhere to higher-quality standards.

Attracting foreign direct investment. To diversify its economy into service-oriented sectors, the government of Mauritius has moved aggressively to attracting foreign investment into IT with the construction of a cyber-city, which became operational in 2004. The increase in FDI has been mainly in the financial intermediation service sector, which is in line with the government's aim to diversify into a more service-oriented economy. Both the firms interviewed in the IT sector were housed in the cybercity. Another such initiative of the government has been to set up the Seafood Hub Mauritius. The emergence of the Seafood Hub is the result of 10 companies involved in fishing activity in the Indian Ocean, storing, preserving, and processing fish products and investing massively in the seafood industry. The government is providing various incentives and facilities, such as trading and warehousing; processing and distributing; reexporting fresh, chilled, and frozen raw or value-added seafood product; assisting with rapid administrative and operational procedures and clearances; and lending to operators who set up fish- and seafood-processing plants and are involved in the export of processed fishery products. The foreign-owned firm in seafood processing is part of Seafood Hub Mauritius.

Although foreign-owned firms originally came to Mauritius to serve the African market and to take advantage of the enabling business environment, firms needed to continuously invest in new technology to meet international standards imposed by the parent company. Moreover, being part of a larger company or group of companies provides the funds necessary to conduct R&D. This investment gives the foreign firms a substantial competitive advantage, since access to local finance is problematic for many of the domestic firms interviewed.

One of the two IT firms deals mainly with data center management and payment exchange, and the firm's main clients are foreign financial institutions. The company is owned by a French firm that has sent over a few of its technical experts to train the firm's staff. The second firm is a U.S.-owned foreign firm that specializes in ICT outsourcing. Of its 370 employees, 160 software engineers are to a large extent expatriate labor because of the shortage of skilled IT professionals available locally. The firm's main clients are financial institutions and retailing businesses, with main markets being the United Kingdom, the United States, and Canada. The capital equipment manufacturer produces galvanized pipes and tubes for industrial and agricultural use. It is a British subsidiary and was located in Mauritius to serve the African market, where it exports to more than 20 countries in the region. Products are all produced under the British standard to conform to its parent company's standards, and technology was brought in to enable the firm to do so. The seafood processor is part of a larger group and is 50 percent owned by a French company. It specializes in frozen fish and lobster handling and has invested in equipment to keep the fish at a certain temperature to ensure that it retains its hydration, firmness, and taste when frozen. The reason for locating operations in Mauritius was the access to the fresh catch and the ability to enter new markets in Europe, including France, Germany, Italy, Portugal, and Spain.

As part of a larger company or a group of companies, these firms were found to have resources and opportunities to invest in R&D. Firms in the

Seafood Hub, for example, that process frozen seafood must adhere to high levels of phytosanitary standards, which require constant testing and R&D. These firms have an export market in Europe and devote a certain share of their revenues to R&D. The firm in Seafood Hub Mauritius also received R&D assistance from its parent company, which has ties with a research institute. This assistance helped the firm gain access to otherwise locally unavailable R&D.

Constraints on enhancing technological competencies. Small firms in the sample were found to be risk averse and financially constrained to explore new markets. The ES regression analysis suggests that small firms had a lower probability of introducing a new product or a new technology. The investments made by firms were correlated to their access to information, new markets, finance, and public support.

Insufficient access to information. Financial constraints aside, small and medium firms also face information gaps inhibiting their ability to keep up with the latest technological developments. Many of the firms are unaware of the merits of increased investment in technology and new, improved production processes, indicating a greater need to sensitize firms, especially small and medium enterprises (SMEs), to the merits of increased investments in new technologies and production processes. The U.K. Manufacturing Advisory Service is one such example that targets information gaps among SMEs, helping raise awareness of possible areas of productivity improvements. Program staff is proactive in helping SMEs understand their needs to identify the right type of intervention.

In Mauritius, various programs exist, such as matching grants for firms that would want to go on marketing trips, participate in trade fairs, and receive consulting services. The firms, however, are often unaware of the organization that oversees these grants or of the eligibility criteria for the grants. Often, this lack of knowledge results in the same set of firms accessing the grants while the others are unaware. A lack of coordination is perceived to exist among the various organizations, such as Enterprise Mauritius or the Small and Medium Enterprises Development Authority. The government is aware of the problem and has proposed a development agency model that could rationalize the various schemes and avoid duplication and possible overlapping functions. The effectiveness of some of these schemes and possible lessons learned are being dealt with in the forthcoming study on technology absorption and skills in Mauritius (World Bank 2011b).

Perceived skill shortages. Both domestic and foreign firms noted issues with respect to labor. In fact, 50 percent of foreign firms in the 2008 ES reported the shortage of skills as a major constraint. All the foreign firms interviewed cited the lack of locally available skilled personnel as a constraint and all have hired as much as 80 percent of their workforce from abroad. The regression analysis suggests that firms providing formal training programs have a significantly higher probability of both introducing a new product (table 2C.12 in annex 2C) and introducing a new technology (table 2C.13 in annex 2C). The relationship may well exist the other way, with firms that introduce new technology being more likely to provide formal training programs to absorb and operate the new technology effectively.

The shortage of skilled labor is further exacerbated by high levels of absenteeism among local employees, leading to firms employing a large amount of expatriate labor from Bangladesh, India, and Sri Lanka. As mentioned earlier, a shortage of adequately trained IT professionals implies the reliance of these firms on expatriate skilled professionals. Furthermore, even in the textile and garment sector, firms are relying on expatriate labor from Bangladesh, India, and Sri Lanka who are regarded as being more productive and cheaper than local labor (World Bank 2010a). An estimated 15,000 textile workers lost their jobs between 2001 and 2005. This number includes a large proportion of women who have weak prospects with respect to retraining, given their gender and age (World Bank 2010a).

The lack of adequate skills in the labor market could potentially discourage entry and new investments and the move into higher-value-added sectors. For Mauritius to move from BPO to knowledge process outsourcing, more advanced skilled labor is needed (World Bank 2010a). This point is particularly important because Mauritius faces tremendous competition from lower-cost centers that are able to cater to client needs because of their access to skilled professionals.

Limited collaboration between industry and research. No collaboration was reported between industry and research. Of the 16 firms interviewed, only 1 firm used the resources of the university to train its workers on AutoCAD (computer-aided design) equipment. Another firm (a jewelry firm) had links with a German research institute through one of its foreign clients because no local expertise was available. The foreign firms all tap into the resources of their parent company to undertake R&D as well as to link up to expertise abroad. The general perception among the firms is that the local research institutions and universities do not conduct any practically relevant research for industry. The cause appears to be a perceived lack of motivation for the research community to engage in intellectual property activities, collaborate with industry, or even aim at securing industry-supported research grants (Kuriakose, Goldberg, and Kaplan 2009). The Mauritian Research Council has a private sector collaborative grant scheme that has been in effect since 1998. To date, only 30 approved projects have benefited from this grant. Analysis suggests that because of the time-intensive application process, these applications are mainly research led and not industry driven and hence are not relevant to industry needs (World Bank 2011b).

Access to finance. The large firms interviewed accessed commercial banks for loans, and in some cases large and medium firms accessed concessional loans from the development bank in Mauritius. Smaller firms were often inhibited from undertaking investments to introduce new products or new technology because of financial constraints. Access to soft loans and other sources of finance was found to increase the ability or willingness of firms to take risks. The inability to invest in R&D is often a result of finance being a key constraint. A small share of firms performed R&D (18 percent), with most of it being financed by internal firm funds and bank loans. In terms of firm characteristics, larger firms were more likely to conduct R&D than smaller and medium firms. The 2008 ES indicates that 24 percent of the firms that do not undertake R&D cite lack of financing as the critical impeding factor.

The case interviews suggest that the lack of adequate finances often means that firms buy secondhand equipment because of their financial constraints. Purchase of secondhand equipment means that the supplier training that generally accompanies the purchase of new equipment is not available to these firms. An important source of technology transfer is thus not available to many firms, particularly smaller and less-capitalized firms.

Summary

While moving away from textiles and sugar, Mauritian firms are moving toward higher-value-added niche products to gain new markets. Exporting to more advanced economies such as the United States and Europe provides firms with the incentives required to invest in new technology and process improvements. Purchase of machinery and equipment continues to be an important source of technology transfer. However, smaller firms with financial constraints tend to buy secondhand machinery and equipment, which implies that these firms are not able to access the training that suppliers of new machinery would give them. A perceptible skill shortage exists, which is alleviated by the ease of hiring expatriate workers. The foreign-owned firms operate in Mauritius to access markets in Africa and in Europe and offer access to additional resources to undertake research as well as transfer of technology. Industry research collaboration is virtually nonexistent, with firms finding local research not industry oriented. Although numerous programs are in place in Mauritius, information dissemination appears to be an issue, with many firms unaware of the programs or the agencies responsible for them.

Lesotho: Textile Sector

The textile sector, which is fully export oriented, is arguably the only sector in Lesotho that has attracted significant amounts of FDI. This section is based on 10 firm-level interviews, the completion of a questionnaire that contained a mix of qualitative and quantitative questions, and plant visits to 10 foreign-owned firms in the country. The formal clothing-manufacturing sector in Lesotho consists of about 43 firms and is said to employ 47,000 people. The total employment of the 10 interviewed firms is 15,940, therefore providing a snapshot of 34 percent of the industry by employment and 25 percent in terms of the firms' population.

Preferential trade access through AGOA and the MFA spurred Taiwanese investment in the Lesotho textile sector in the early 2000s. Clothing exports jumped from an insignificant US$6.7 million in 1995 to US$455.9 million in 2004. With the end of the MFA in 2004, however, many of these manufacturers closed their factories, dropping exports to US$339.7 million by 2008. Over the last few years, the remaining Taiwanese-owned clothing manufacturers in Lesotho have been joined by a growing number of South African–owned clothing manufacturers looking to escape South Africa's higher-cost operating environment. Lesotho's membership in the Southern African Customs Union provides duty-free access to South Africa, and these South African–owned firms mainly produce for South African retailers. This pattern is evident in the export destinations of the two types of firms, with 94 percent of the Taiwanese output going to the United States and almost all the South African output going to South Africa.

The Lesotho clothing industry remains highly vulnerable to intensifying cost competition from locations such as Cambodia and Vietnam. The appreciation of the loti (which is pegged to the South African rand) against the dollar has not helped the situation. The Duty Credit Certificate Scheme (DCCS), which effectively subsidizes exports in the amount of 14 percent to 25 percent of sales, is also being phased out. These considerations raise serious questions as to the sustainability of the Lesotho clothing industry. Because Taiwanese FDI is contingent on the export schemes, whether Lesotho is capable of (or willing to engage in) technology upgrading and dynamic growth is uncertain.

The majority of the Taiwanese firms interviewed were established prior to 2000 to get preferential access to the United States through AGOA, and the South African firms joined thereafter to escape the higher-cost operating environment in South Africa. The different investment motivations of South African and Taiwanese firms are displayed in figure 2.4. All the firms are still owned by residents of their respective countries except for one of the Taiwanese firms, where the owner took Lesotho citizenship. The Lesotho-based operations are larger than those in South Africa, with 80 percent of the Taiwanese firms and 40 percent of South African firms in Lesotho having more than 800 employees, compared with 15.5 percent for operations in South Africa.

The Taiwanese-owned firms' average capital expenditure of 9.3 percent is radically different from the South African–owned firms' figure of 1.6 percent. The Taiwanese-owned firm average is, however, skewed by one firm spending the equivalent of 30 percent of its sales on capital equipment in 2009. Excluding this firm, the average for Taiwanese-owned firms would be 4.2 percent. The South African–owned firms'

Figure 2.4   Clothing Manufacturers' Reasons for Investing in Lesotho

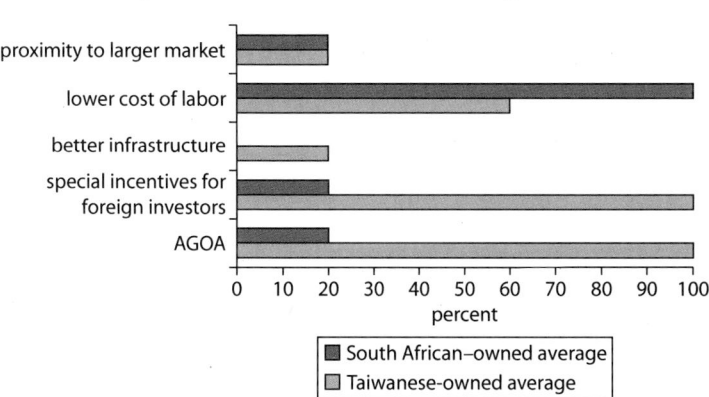

Source: Case interviews.

capital expenditure is on par with the South African firm-level average. Eight of the 10 firms function as cut-make-and-trim garment manufacturers, one of the Taiwanese firms functions as a full clothing manufacturer, and one of the South African firms is both a cut-make-and-trim and full manufacturing firm.

Technology Absorption in Textile Firms in Lesotho

Almost all of the Taiwanese firms and all of the South African firms interviewed have their head offices in their home country. Four of the five Taiwanese firms have their head offices in Taiwan, China, and all of the South African–owned firms have their head offices in South Africa. These head offices are responsible for marketing, liaising with customers, obtaining orders from customers, designing, managing logistics, and procuring raw materials from suppliers.

Eight of the 10 firms are cut-make-and-trim garment manufacturers; that is, they do not procure their own fabrics and trims and have no input into the design process relating to the garments they assemble. The product-type profile of the South African–owned firms is slightly more diverse than that of the Taiwanese, with 40 percent of the garments categorized as moderate or constructed (more tailoring than a simple item). The end-market segmentation explains some of the important differences between Taiwanese- and South African–owned firms with regard to export product profile (Morris, Staritz, and Barnes 2011). Exporting to the U.S. market, Taiwanese-owned firms produce basic clothing products that are exported in bulk. South African–owned firms focus on shorter-run products, which are more time sensitive and have greater fashion content. For both the South African and the Taiwanese firms, the Lesotho plants are used almost exclusively for manufacturing. This limits the opportunity to absorb any new product or non-manufacturing-related technologies, such as advanced IT interfaces with retailers and fabric suppliers. The innovation potential for this value-chain structure is therefore limited.

The low-technology production strategy is also evident in the productivity and performance indicators. Both the South African and Taiwanese firms in Lesotho had high internal rework rates, and the South African firms had high customer returns.[20] These rates suggest major operational deficiencies and an inability or unwillingness to embrace organizational technologies, such as total quality management, which build quality into production processes. Moreover, both types of firms displayed operational inflexibility, measured by production downtime due to style changeovers. The mean age of capital equipment for both types was over seven years, and operational reliability was fairly low (in terms of supplier and customer deliveries and production downtime). This situation was compounded by high turnover rates of 19.1 percent for the Taiwanese and 12.8 percent for the South African firms. These high turnover rates reduce the incentive to provide training, further worsening the skill shortage.

Constraints on enhancing technological competencies. As with investment in human capital, process and product innovation have been limited. Since their inception, only two Taiwanese firms and one South African firm have conducted any product innovation. The same statistics are true for process innovation, with the South African firm's investment being the only one that consisted of more than general machinery upgrading. The firms attributed this fact to their value-chain configuration: Lesotho merely functions as a basic manufacturing site for innovation originating at the headquarters.

Skill shortage. The skill shortage in Lesotho is particularly onerous for technology absorption. Nine of the 10 firms struggled to find skilled labor and highlighted the very basic level of instruction at the Lesotho training schools for machinists. Although all claimed to conduct internal training, total spending on training is low: 0.38 percent of total remuneration for Taiwanese firms, and 1.03 percent for South African firms. As seen in figure 2.5, this expenditure is much lower than in neighboring South Africa. Furthermore, the Taiwanese firms employ almost no Lesotho citizens in management positions. Although top management in the South African firms is still predominantly South African, they do employ more Lesotho citizens in middle- and lower-management levels.

Other barriers to increased investment in technology absorption. The interviewed firms cited a number of specific barriers to further investment, notably logistics (customs and land shipment costs of imported materials from the East), start-up finance, and a severe shortage of skills. The South African firms were predominantly concerned about skills, however, and the Taiwanese about costs. This distinction between the two types of firms has important implications for technology absorption. South African–owned firms take skill development and operational upgrading more seriously because these are a condition for the transfer of more work from their South African to their Lesotho operations. In contrast,

Figure 2.5   Training Investment by Lesotho Clothing Manufacturers

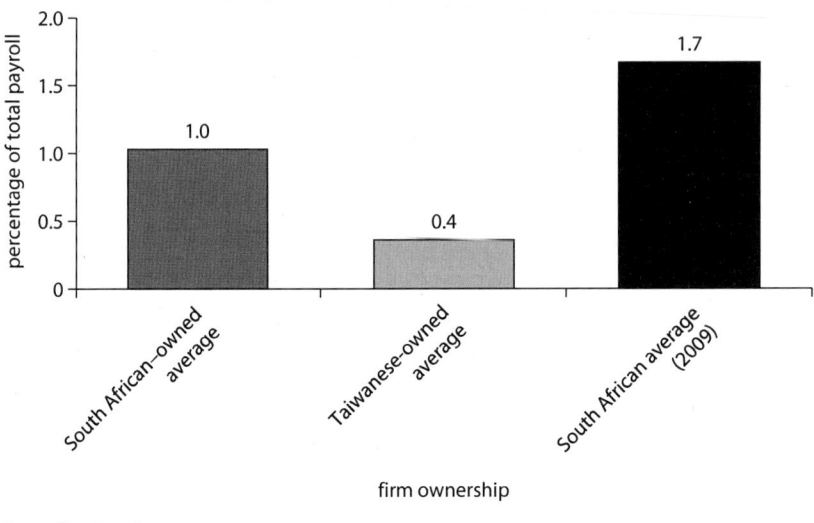

Source: Case interviews.

the long supply chains of the Taiwanese firms necessitate cost-control prioritization.

Future prospects. With respect to policy support, the firms are critical of the pegging of the loti to the South African rand, especially in the wake of the recent appreciation against the dollar. Three of the Taiwanese firms noted that the peg, together with the impending phaseout of the DCCS, would result in their closure. Four of the Taiwanese firms and one South African firm are members of the Lesotho Textile Exporters Association, which focuses on lobbying the Lesotho government and the U.S. Congress for continuation of the DCCS and AGOA. Apart from this, few of the Taiwanese firms are taking other measures to offset the probable loss of these benefits in the near future. Like the Taiwanese firms, the South African firms consider the DCCS important. Unlike the Taiwanese firms, however, the South African firms consider control of the labor market and the provision of access to credit critical areas for policy. With respect to other government interventions, all firms have received a limited amount of support from the Lesotho National Development Corporation, including subsidized factory rentals, support for labor recruitment and industrial relations issues, and establishment of basic skill development centers for machinists in Maseru and Maputsoe.

Summary

This snapshot of FDI in the Lesotho textile industry is revealing. The sector is facing severe constraints, but in many ways this situation is a function of the type of FDI itself. Both the South African and the Taiwanese plants have been set up as low-cost assembly operations in support of well-established, higher-value-adding design-and-development facilities located elsewhere. This offshore siting has been done to take advantage of both preferential market access into the United States (for Taiwanese-owned firms) and fewer labor regulations (for South African–owned firms). The firms, particularly the Taiwanese, are then forced into narrow value-chain positions that make them extremely vulnerable to any factor cost changes in their operating environment. Where the South African–owned plants are different is in their willingness to develop and promote Lesotho nationals into supervisory and more senior management positions. A notable gap in the provision of training within Lesotho relates to the lack of any formal relationship between the clothing industry and Lesotho's university. Firms indicated that they had never participated in research completed by the local university or been engaged to participate in any potential university-industry projects. This lack of engagement represents a major institutional failure in terms of potential technology absorption opportunities.

Conclusion

Many of the industries in southern Africa and their exports of manufactures rely on imported technologies. Almost all sophisticated capital equipment is imported, and technology agreements with foreign firms are widespread. However, exports from these countries tend to be concentrated in the lower-value-added products: textiles in Mauritius and Lesotho, agroprocessing in Namibia. Even in South Africa, many of the chemical and auto component products tend to be concentrated at the lower-value end of the scale.

Key findings from the NIS and the ES analyses suggest the importance of R&D and human resources for both South Africa and Mauritius. R&D has the largest (positive) effect on the probability of innovating. Evidence from case studies also suggests that successful exporting firms devoted resources to R&D that enabled the firms to compete globally. Although the ES instruments do not distinguish between new-to-firm, new-to-country, and new-to-world innovations, the NIS has a measure of new-to-firm versus new-to-market innovations. For comparison of new-to-market and new-to-firm product innovation, results suggest that R&D, public support (despite a low number of firms that receive it), acquisition of external knowledge, and registration of design were significant. Thus, a need exists for a more effective research and innovation policy framework that enables leveraging of public financial support for R&D and innovative activity. The government can stimulate R&D investments and innovation in the private sector through effective public financial support mechanisms. The case interviews suggest the need for policies in South Africa to be directed toward using accumulated expertise and technological sophistication to encourage greater spillover into areas beyond mining. Of particular concern is that very few firms interviewed regard government policy supports for technology development to be effective: the firms either are unaware of the existing programs to support R&D and technology or view them as ineffective and bureaucratically cumbersome.

The NIS instrument allowed differentiating between product and process innovation for the entire sample; product and process innovation for the sample of innovative firms only; and innovation intensity, measured by R&D participation and by whether product innovations are new to the market or new to the firm. For the full sample of firms, the only significant findings are that medium firms and IPR were significant determinants of new product and process introductions, though for new processes, financial constraints also became significant when the specification of the model changed to incorporate foreign subsidiaries relative to domestic subsidiaries. The results for R&D suggest that manufacturing firms are significant compared with services; exports, cooperation on innovation, and a variety of IPR variables were significantly related to the probability of undertaking R&D activity. The case interviews also suggest that firms in the mining and mining equipment sector that are close to the technology frontier all invest in R&D. These firms have also in the past relied to a large extent on the research capabilities that were available in Mintek (the mining research institute) or universities and the CSIR. These capabilities in the research institutes have been on the decline, which has a direct impact on the research and innovative capabilities of the South African mining firms that are losing out to their global competitors, especially from Australia.

For the innovator-only subsample of firms, however, many more variables become significant for new product introductions, including firm size, foreign subsidiaries, exports, IPR, R&D, knowledge intensity, and cooperation on innovation. For the introduction of a new process innovation, an interesting finding is that firm size and sector were not significant, suggesting that process innovations are more diffuse across sectors and firm sizes. Having partnerships with foreign firms or being subsidiaries of foreign firms does indeed lead to increased R&D by firms, so that many at times tap into the research capabilities available either in the firm's head office or in the research institutes in the parent company's country. This propensity was evident among firms interviewed in Mauritius and Namibia.

Foreign firms have the potential to act as important conduits for technology transfer and "upgrading" in host countries. Through demonstration effects, vertical links, staff turnover, and competitive pressure, FDI can help accelerate countries' progress along the development path. However, in addition to making sure that the incentives for these foreign firms are in place, the one issue area that *does* require a much longer outlay to benefit fully from FDI is that of worker skills. FDI (proxied by majority foreign ownership) is not unambiguously positive or consistently statistically significantly associated with the probability of introducing product or process innovations in both samples for South Africa and Mauritius. This finding could imply that if foreign firms are not investing in the creation of new technology, then their interest in these countries is market seeking rather than efficiency seeking. MNCs are investing in these countries to get access to local markets and the subcontinent of Africa— rather than to develop new technologies. This conclusion is definitely the case for foreign investors in Lesotho, who have located their operations in Lesotho to benefit from AGOA and to gain access to U.S. markets. Such is the case for the Taiwanese-owned firms; the South African–owned firms located in Lesotho escape the higher-cost operating environment in South Africa. A need thus exists for countries in southern Africa to develop the capability to become a destination for efficiency-seeking FDI at some point in the near future. Furthermore, the positive and significant relation between joint ventures and introducing a new product or a new technology in South Africa indicates the importance of foreign investment to technology transfer.

The ES analysis highlights a few key issues that pose as the key deterrents to foreign investment: shortages of skilled labor, labor law inflexibility, regulatory and political uncertainty (including over BEE laws), red tape, corruption and crime, and exchange controls. Other noted areas are exchange rate instability in Lesotho, Namibia, and South Africa; electricity in Lesotho, Mauritius, and South Africa; and tax rates in Namibia. All these factors can be affected by or ameliorated through policy intervention. Hence, if countries wish to attract more FDI, these are the issues that governments will need to address.

Skill shortages are cited as a constraint in most industries across all the countries. Moreover, respondents in South Africa overwhelmingly view the present system of training as nonoptimal. For the South African sample, worker training and the share of skilled workers in the firm do not have large or statistically significant effects (either positive or negative) in the most elaborate specifications of the models. This finding suggests that the training regimen in South Africa is currently not effective enough for increasing the absorptive capacity of workers. Possible reasons could include, for example, the ineffectiveness of the SETA system with respect to facilitating skills and learning of sufficient quality. This reason is corroborated by the case interviews, in which all firms interviewed found the present SETA system ineffective for firm needs. Furthermore, in the NIS, the lack of qualified personnel was the second highest constraint reported by firms. This skill shortage is further aggravated in Namibia and South Africa by the difficulty of acquiring the necessary skills from abroad, which is in sharp contrast to Mauritius, where skills are easily sourced from overseas to overcome the skill gap locally.

Other constraints faced by firms interviewed included the insufficient and often inefficient public support schemes available to provide incentives for investment in R&D, increase industry research collaboration, and increase collaboration across country boundaries between firms, researchers, and academia.

Though policy needs to be directed at addressing skill shortages in these sectors, both in the short and the medium to long term, addressing the barriers to effective R&D collaboration and exploiting expertise in clusters of excellence to bring that expertise to new areas in the economy are also critical. This would entail more incentives for R&D investments and increased industry-research collaboration. All the countries are committed to greater diversification, especially into more high-value-added sectors, thereby putting technology absorption at the crux of their strategies. Effective public support plays a vital role in this endeavor, and chapter 3 proposes possible policy solutions that could alleviate some of the constraints faced by firms that now hinder greater levels of technology absorption.

Annex 2A: Data Supporting the Case Study Analyses

This annex provides figures and tables that support the case study analyses.

Table 2A.1   South Africa's Top 10 Automotive Component Export Destinations, by Country, 2004–08

percentage of total

Country	2004	2005	2006	2007	2008
Germany	35.9	33.3	29.3	28.5	33.1
Spain	8.3	9.7	11.5	10.4	9.8
United Kingdom	9.0	9.6	9.3	7.8	8.4
United States	7.2	9.5	9.7	8.0	7.7
France	7.9	6.6	7.0	7.8	5.8
Belgium	3.1	3.1	4.4	5.0	5.2
Poland	0.6	1.5	2.9	2.7	3.2
Netherlands	2.8	1.9	1.8	2.4	2.3
Brazil	0.5	1.0	0.8	0.9	1.6
Japan	2.3	1.2	2.4	2.0	1.6
Other	22.4	22.6	20.9	24.5	21.3

Source: AIEC 2009, 32.

Table 2A.2   South Africa's Automotive Component Export Destinations, by Region, 2004–08

Region	2004 (rand, millions)	2005 (rand, millions)	2006 (rand, millions)	2007 (rand, millions)	2008 (rand, millions)	2008 (%)
European Union	19,973	21,706	27,313	32,509	40,680	41
NAFTA (North American Free Trade Agreement)	4,449	3,569	6,170	8,820	18,304	18
Africa	2,860	3,454	5,402	6,929	12,100	13
SADC	2,167	2,268	2,784	3,860	6,066	6
Mercosur (Southern Cone Common Market)	128	237	275	422	785	1
Other	29,577	31,234	41,944	52,540	77,935	21

Source: AIEC 2009, 5–9.

Table 2A.3   South Africa's Top 10 Automotive Component Exports, by Component Category, 2004–08

Component category	2004 (rand, millions, current prices)	2005 (rand, millions, current prices)	2006 (rand, millions, current prices)	2007 (rand, millions, current prices)	2008 (rand, millions, current prices)	Percent of 2008 total
Total	21,733	23,277	30,052	39,106	44,055	n.a.
Catalytic converters	8,288	9,934	15,810	21,683	24,245	55
Leather seats	3,113	2,693	2,549	2,760	3,282	7.4
Silencers/exhausts	407	492	407	1,705	1,900	4.3
Engine parts	894	1,000	984	1,092	1,888	4.3
Tires	1,285	1,183	1,220	1,196	1,670	3.8
Engines	701	781	1,216	1,060	1,050	2.4
Automotive tooling	383	332	272	520	800	1.8
Transmission shafts	332	553	351	556	782	1.8
Road wheels and parts	753	738	681	772	691	1.6
Gauges/instruments	142	161	184	248	582	1.3

Source: AIEC 2009, 21.
Note: n.a. = not applicable.

Table 2A.4   South Africa's Top 10 Export Destinations for Capital Goods, January–June 2009

Country	Rank	Value (US$)	Percentage of total
Zambia	1	973,910,045	10.37
Zimbabwe	2	844,610,729	8.99
Mozambique	3	822,540,428	8.76
Congo, Dem. Rep.	4	495,877,908	5.28
Angola	5	439,473,404	4.68
United States	6	436,203,767	4.64
Nigeria	7	407,975,237	4.34
United Kingdom	8	315,089,622	3.35
Tanzania	9	297,695,436	3.17
Malawi	10	289,712,923	3.09
Total top 10		5,323,084,000	56.67
Africa share of top 10		4,571,792,000	86.00

Source: SACEEC Business Plan 2009–10 (provided by interview respondent).

Table 2A.5   South Africa's Mining Equipment Export Destinations, 2005–09

Percentage of total exports

Destination	2005	2006	2007	2008	2009
World (US$, millions)	3,211	4,639	6,077	6,617	4,027
Germany	15	13	12	16	14
Spain	11	10	9	8	5
France	8	8	8	7	5
United Kingdom	8	9	7	6	5
United States	9	6	6	6	5
Sub-Saharan Africa	25	22	25	29	38

Source: SACEEC Business Plan 2009–10 (provided by interview respondent).

Table 2A.6   South Africa's Mining Equipment Exports to Sub-Saharan Africa, 2005–09

Percentage of total exports

Destination	2005	2006	2007	2008	2009
Mining Equipment Exports to Sub-Saharan Africa (US$, thousands)	786.9	1,025.8	1,494.1	1,936.0	1,542.7
Zambia	18	20	21	16	16
Zimbabwe	12	12	11	14	14
Mozambique	11	12	8	9	13
Nigeria	6	6	8	7	10
Angola	9	8	10	9	9

Source: SACEEC Business Plan 2009–10 (provided by interview respondent).

Table 2A.7 South Africa's Principal Chemical Exports, 2005–09

| ISIC | Description | Exports (rand, thousands) | | | | | Percentage of total manufactured exports |
		2005	2006	2007	2008	2009	2009
3341	Basic chemicals	13,725,477	13,983,298	15,737,753	26,074,764	13,477,640	5.6
3320	Petroleum refineries	10,049,918	9,236,672	8,876,270	15,011,714	9,824,360	4.0
3343	Primary plastics and synthetic rubber	2,695,103	2,734,396	2,818,794	5,246,219	3,709,267	1.5
3359	Other chemical products	2,382,000	2,856,963	3,990,259	4,487,093	2,927,832	1.2
3354	Soap and detergents	1,400,259	1,785,168	1,804,856	2,701,219	2,400,921	1.0
3380	Plastic products	1,520,795	1,640,997	2,064,277	2,841,825	1,956,012	8.0
3342	Fertilizers and nitrogen compounds	1,258,001	1,190,179	1,420,993	2,853,884	1,567,352	0.6
3353	Pharmaceuticals	957,051	1,037,688	1,322,730	1,838,900	1,564,173	0.6
3352	Paints and varnishes	868,728	706,267	1,339,075	1,919,470	1,397,443	0.6
3371	Rubber tires and tubes	1,344,491	1,436,443	1,483,207	2,070,917	1,361,377	0.6
3351	Pesticides	904,582	893,992	971,809	1,008,507	1,048,610	0.4
3379	Other rubber products	411,388	495,940	546,822	670,271	535,559	0.2
3360	Manmade fibers	603,779	586,171	691,261	568,702	234,117	0.1
	Total	212,697,831	252,514,261	316,192,964	418,059,763	242,800,566	17.2

Source: Department of Trade and Industry (DTI) database, http://www.thedti.gov.za.
Note: ISIC = International Standard Industrial Classification.

Table 2A.8 Top 16 Export Destinations for South African Chemicals

Rand, thousands, 2011

Country	2009 Exports (US$)	2009 Share of top exports (%)	2008 Exports (US$)	2008 Share of top exports (%)	2007 Exports (US$)	2007 Share of top exports (%)	2006 Exports (US$)	2006 Share of top exports (%)
Angola	1,317,817	8.0	1,267,919	8.2	777,784	6.9	675,309	6.7
Australia	461,467	2.8	474,606	3.1	446,299	4.0	347,465	3.4
Canada	479,085	2.9	537,124	3.5	393,049	3.5	397,645	3.9
China	263,779	1.6	280,141	1.8	242,695	2.2	233,339	2.3
Germany	1,280,805	7.8	1,282,404	8.3	941,864	8.4	708,249	7.0
Italy	877,038	5.3	976,595	6.3	792,352	7.0	562,936	5.5
Japan	939,500	5.7	1,011,988	6.5	838,758	7.5	808,856	8.0
Kenya	329,292	2.0	377,676	2.4	248,515	2.2	232,948	2.3
Mozambique	1,582,493	9.6	1,641,703	10.6	1,167,500	10.4	1,015,245	10.0
Netherlands	1,034,553	6.3	1,300,779	8.4	1,040,824	9.3	1,073,555	10.6
Spain	498,460	3.0	782,496	5.1	583,147	5.2	352,971	3.5
Sweden	756,151	4.6	708,363	4.6	516,423	4.6	371,565	3.7
United Kingdom	2,135,835	13.0	2,134,388	13.8	1,742,444	15.5	1,345,703	13.3
United States	1,158,146	7.1	1,057,741	6.8	969,711	8.6	1,061,912	10.5
Zambia	773,735	4.7	704,896	4.6	306,104	2.7	548,807	5.4
Zimbabwe	2,511,547	15.3	929,260	6.0	231,734	2.1	414,314	4.1

Source: DTI database, http://www.thedti.gov.za.

Table 2A.9   Percentage Composition of Manufactured Products, Namibia

Year	Meat processing	Fish processing	Food and beverages	Other manufacturing	Total manufacturing
1995	8.4	26.7	36.9	28.0	100.0
1996	11.1	11.4	48.2	29.2	100.0
1997	7.0	17.0	46.8	29.2	100.0
1998	6.4	26.6	44.7	22.3	100.0
1999	6.7	21.7	48.9	22.7	100.0
2000	5.1	23.1	46.0	25.8	100.0
2001	5.5	19.0	46.7	28.9	100.0
2002	4.3	21.3	45.8	28.6	100.0
2003	3.6	22.6	42.6	31.1	100.0
2004	3.2	18.8	42.3	35.9	100.0
2005	3.0	11.5	43.7	41.8	100.0
2006	2.5	15.3	47.2	35.0	100.0
Average	5.8	20.0	44.8	29.4	100.0

Sources: Central Bureau of Statistics, National Planning Commission, Windhoek, Namibia; Kadhikwa and Ndalikokule 2007.

Annex 2B: Survey Instrument

Basic Firm Characteristics

Date and time	
Name of firm	
Industry	
Year started	
Number of employees?	
Part of larger firm?	
% foreign owned	
% state owned	
Any R&D?	
Formal training program?	

Export Activity

A. The first set of questions is to establish the importance of exports to the firm; is it export oriented and why/why not; what are the likely future changes?

What items do you export and to where?

What were the reasons for you to start exporting? Did excess capacity or insufficient domestic demand play a role?

What place do exports have in this vision of the firm? How important are they to the overall strategy of the business?

How have the products exported and the markets exported to been changing over time?

What markets do you export to?

		(Yes/No)
A	SADC	
B	Other countries in Africa	
C	South Asia	
D	East Asia	
E	United States	
F	Europe	

Did you have to make changes to your product to be able to export to these markets?

What is the share (%) of new product that you have introduced in the past three years in your total exports in FY2007?

What else affected your decision to enter this market?

Did you have personal or family connections to anyone who helped you enter this market? (Yes/No)

How do you forecast your exporting will change in the next 5–10 years? Do you anticipate that you will have a larger/smaller share of the global market than currently? Are you planning to occupy the same market value niche in the future or do you anticipate that that will change? Are you planning to enter new global markets?

Is your firm more or less export oriented than your competitor firms in the domestic market? To what do you ascribe the difference, if any?

B. The next set of questions relates to the constraints and difficulties of exporting.

What are the main costs entailed in exporting? How do you finance these costs?

What channels do you use to access your main export markets? (own office abroad, foreign intermediaries, direct to end user)

Who are your primary competitors in your main export markets?

What is your competitive edge over these competitors?

How do wage costs in your firm compare with your competitors in global markets? (Think of a similar-sized firm, producing a similar product for a similar market segment.)

How does the productivity of workers in your firm compare with your competitors in global markets? (Think of a similar-sized firm, producing a similar product for a similar market segment.)

How significant are infrastructural constraints in retarding exports? Transport to the ports; the ports; shipping? How do the constraints compare with your competitors' constraints?

Are there tariff duties on inputs that you use for exports? How significant are they—both in terms of cost and delay—and to what extent do they constrain your export activity?

What factors constrain your level of export activity (high cost of labor, finance, exchange rate, bottlenecks)? Please elaborate on these constraints.

What are the really key constraints such that if they were removed, the company would significantly expand exports? (rank them)

What are the most important lessons that you have learned through your company's export experiences?

Technology and Innovation

C. The focus of the next set of questions is to explore the importance of innovation/technology to the company.

How important is technological innovation in the strategy of your company?

What are your current innovation priorities?

To what extent does the company have sufficient resources to invest in these innovation activities?

What have been your most important innovation successes? How much financial value did these create for the firm? What were the keys to these successes?

How many new or significantly improved products did you introduce?

What is the new or significantly improved product(s) that the firm introduced over the last three years?

How is your product different from the most similar product sold by (a) your firm and (b) other firms in South Africa?

What was the most important objective of introducing this new or significantly improved product? (Please choose one of the following options.)

Diversifying your firm's product mix for the domestic market	1
Diversifying your firm's product mix for foreign markets	2
Increasing domestic sales in market segments in which you were already operating	3
Increasing foreign sales in countries in which you were already operating	4
Entering new foreign markets	5
Keeping up with innovations introduced by domestic or foreign competitors	6
Other (_____)	7

For the most important new or significantly improved product, how much did you spend on introducing it, including research and development costs, licensing costs, and market research? (in South African rand)

For the most important new or significantly improved product, did you have to purchase capital equipment to introduce the new product? (Yes/No)

Did your workers need to be retrained to use the capital equipment that you purchased? (Yes/No)

What was the most important source of training for the capital equipment that you purchased? (Please choose one of the following options.)

Internal training by your own employees	1
Equipment supplier	2
External consultant	3
Technical college or other external training program	4
Other (_____)	5

Was the capital equipment secondhand? (Yes/No)

Where was the capital equipment made?

For the most important new or significantly improved product, did you license the product from another institution or enterprise, including a parent company? (Yes/No)

From whom did you license this product?

Parent company	1
Foreign institution or enterprise	2
Other domestic institution or enterprise	3

Did you need to change suppliers when you introduced the new product?

If yes, what was the country of origin of the new suppliers?

Did you have to help any of your old suppliers to adapt their inputs to meet new specifications required to produce the new product? (Yes/No)

If you had to hire new personnel, did you hire any foreign personnel? (Yes/No)

If yes, please list the nationality of the newly hired personnel.

What has been your most costly innovation failure? Could you briefly describe the circumstances around this?

How does the company protect its intellectual property? Is the new or significantly improved product protected in any of the following ways?

		(Yes/No)
A	Registered trademarks	
B	Patent in South Africa	

C	Patent abroad	
D	Confidentiality agreement	
E	Strategically (e.g., through secrecy or lead time)	

R&D

Do you expect to spend resources in developing new or significantly improved products in the next three fiscal years? (Yes/No)

If no, what are the reasons for not undertaking innovative activity?

		Yes/No
A	Do not need to	
B	Cost of developing products is too high	
C	Do not have qualified personnel	
D	Lack of internal funds	
E	Lack of external financing	
F	Cannot find suitable cooperative partner	
G	Market is dominated by established enterprises	

In the last fiscal year, did the establishment spend resources on R&D activities, including R&D subcontracted to other companies?

Yes	1
No	2
Don't know	−7

In the last fiscal year, how often did the establishment spend resources on R&D activities performed within the establishment or subcontracted to other companies?

	ZAR

In the last fiscal year, what percentage of funds that financed those R&D activities came from the following sources?

		Percent
A	Internal funds/retained earnings	
B	Banks	
C	Universities	
D	Venture capital	
E	Federal and state government	
F	Other	

In the last fiscal year, what were the reasons why this establishment did NOT spend on R&D activities?

	Yes	No
Not interested	1	2
Lack of funds	1	2
Lack of skills available	1	2
No good property rights	1	2
Other (please specify)		

D. **The focus in the next set of questions is to explore how important product and process innovation in the firm is to success in export markets.**

Were these products introduced in the home market or in export markets?

Is this product new to the firm, new to the country, new to the world?

Did you receive external assistance in developing the new product from any of the following?

		(Yes/No)
A	Domestic research and development institute	
B	Foreign research and development institute	
C	Domestic university	
D	Foreign university	
E	Parent company	
F	Equipment suppliers	
G	Suppliers of intermediate inputs	
H	Other domestic enterprise	
I	Other foreign enterprises	

Ask the above three questions in relation to process innovation.

To what extent would you say that technological innovation is at the center of the export strategy?

To what extent is your success in export markets dependent on your company's technology innovation capability?

E. The next set of questions is to establish if and how exporting (as opposed to other sources) enhances technological advance.

Has your experience in export markets had any influence on your company's innovation capability? (Do any of your foreign clients supply you with information, knowledge, or technology which enhances your own innovation activities? Have you learned from competitors in global markets?)

What are the sources of your technological advance?

		(Yes/No)
A	R&D	
B	Technology licensing from abroad	
C	Importation of capital equipment	
D	Training that accompanies importation of capital equipment	
E	Foreign investors	
F	Foreign customers	

What are your most important technological resources?

		(Yes/No)
A	Machinery and equipment	
B	Intellectual property	
C	Development capabilities	

D	Universities	
E	Technology licensed from abroad	

Future Challenges and Constraints

F. This set of questions is to gauge future obstacles to the firm.

What do you see as the major challenges facing the firm in general (competitive threats, protectionism, lack of availability of skills, etc.)?

Do you find it difficult to source skilled labor? Does this difficulty adversely affect (a) exports or (b) technological development?

How much training does your firm undertake as a share of turnover? Has this been increasing/decreasing over time?

To what extent do you use the SETAs, and how effective are they in respect of the needs of your firm?

What constrains and what enhances your willingness to spend on training?

How will the company get around the constraints that it faces? In particular, what role will technology and innovation—broadly understood as including marketing and new organizational forms—play in the future in this process?

Policy Environment

G. This set of questions relates to specific policies in respect of (a) exports and (b) technology.

Have any government programs helped you in your export/innovation activity?

Has the export council and/or trade association for your sector helped you in your export activity?

Have any government programs helped you in enhancing your technology?

Are you aware of or have you made use of SPII, the Innovation Fund, or any other government supports in developing your technology?

Are there any aspects of public policy that negatively affect your ability to compete effectively in export markets?

To what extent is your company's innovation activity supported by public policy? (If positive, what are the most supportive aspects of public policy and how have they enhanced your innovation activity?)

H. This set of questions aims to explore what policy supports competitors in other countries receive and what respondents would propose, in relation to both technology and exporting.

Do your primary competitors in export markets benefit from more or less effective state supports? Please elaborate.

Looking forward, if you had to name just two policies that would help you (a) enhance your exports or (b) enhance your technological activities, what would they be?

Annex 2C: Empirical Analysis Using the NIS and the ES Data

This book uses a variety of surveys and case studies to shed light on the various channels of technology absorption. In this chapter, the findings of the econometric analyses are used to complement some of the findings arrived at from the case studies. The econometric analyses for technology absorption in this chapter use two data sources: the South African National Innovation Survey of 2005, conducted by the Human Sciences Research Council for the Department of Science and Technology, and the World Bank Enterprise Surveys for Lesotho, Mauritius, Namibia, and South Africa.

Review of the Empirical Literature

Using the firm-level, cross-sectoral, and cross-country data set (World Bank Enterprise Surveys), Almeida and Fernandes (2008) provide evidence of the importance of technology transfers through trade, FDI, and licenses for technological innovations for 43 developing countries. Using the ES data for countries in Eastern Europe and Central Asia, Goldberg and others (2008) find that the introduction of new products and processes is positively correlated with an exporter dummy, export as a percentage of sales, majority foreign owned, percentage sales to MNCs, joint venture with MNCs, size, R&D expenditures, and training.

Since the advent of the European Union Community Innovation Survey (CIS), the long-studied relationship between R&D, innovation, and productivity has been revisited using the additional information on the outputs and the modalities of innovation activities contained in the CIS. Crépon, Duguet, and Mairesse (1998) (referred to as CDM in this annex) estimate a model composed of three equations: (a) an equation explaining the amount of R&D; (b) an innovation output equation where R&D appears as an input (CDM had two alternative measures of innovation output: the number of patents and categorical data on the share of innovative sales); and (c) a productivity equation, in which innovation output appears as an explanatory variable. The CDM model has been estimated for a number of countries individually: Chile, China, Estonia, France, Germany, Italy, the Netherlands, Portugal, the Russian Federation, Scandinavia, and Spain. It has also been run with the same specification on four countries: France, Germany, Spain, and the United Kingdom. In the French CIS, Mairesse and Mohnen (2005) find that R&D is positively correlated with all measures of innovation output and that innovation is generally more sensitive to R&D in the low-tech sectors than in the high-tech sectors. They explicitly treat R&D as endogenous and account for the selection of R&D-performing firms. They argue that common causal factors of R&D and innovation do not bias their R&D coefficient estimates, to the extent that the explanatory variables in the R&D equation are indeed exogenous. Further, Garcia and Mohnen (2010) quantify the effect of public support for innovation on innovation inputs and outputs in Austria.

Empirical Analysis

The analysis characterizes the relationship between a firm's probabilities of introducing any of the six dimensions of technology absorption (listed below) and the underlying determinants of absorption outcomes, by estimating a technology absorption production function as shown below:

Technology Absorption = f (K, L, OPEN, IP, R&D, HC, IC, PS),

where K is access to finance; L represents firm employment; OPEN represents the openness of the firm to international trade and knowledge from abroad; IP is the ownership of intellectual property (patents and know-how); R&D refers to whether the firm conducts research and development; HC refers to human capital; IC is a catchall variable for the investment climate; and PS is public support for technology absorption. In this formulation, the determinants can be seen as inputs into this production function. All regressions control for the firm's sector and size as measured by total employment.

The following dimensions of technology absorption in a firm are used:

1. Introduction of a new product.[21]
2. Introduction of a new process.
3. Introduction of new knowledge management systems, a new organizational system, or a new management structure (hence "organizational change").
4. Changes to the design of packaging, sales, or distribution systems (hence "marketing change").
5. Investment in R&D. It is also used as a dependent variable to approximate the output of technology absorption (see further explanation below).
6. New to the market versus new to the firm (see definitions below).

To identify the determinants of technology absorption, the econometric analyses estimated correlations (using multivariate regressions) between the above-mentioned six measures of technology absorption and a variety of explanatory variables that include firm employment, R&D, firm export status, foreign ownership or FDI status, tertiary education, subsidiary status, financial constraints, an aggregate IPR indicator or patents in South Africa, application for a patent outside South Africa, registration of an industrial design, and registration of a trademark.

The term *correlation* is used in reference to the relationship of technology absorption dimensions and the determinants or inputs into the production function, since the direction of causality between the dimensions and the determinants cannot be identified. A positive correlation of technology with export status cannot suggest if the technological capacity causes a firm to become an exporter or, alternatively, exporters increase their technological capacity because of market requirements. Attempts were made during the interviews (case studies) to identify the sequencing by asking whether firms acquired technological capacity before becoming exporters or if they decided to export because they already possessed the capacity. The findings are reported in chapter 2.

Table 2C.1   Introduction of a New Product and New Process in the Full Sample

Variable	New product	New product	New process	New process
Micro	0.296552	0.307636	−0.42442	0.307636
Medium	0.95114454***	0.96026095***	0.62663033*	0.96026095***
Large	0.61307899*	0.572383	0.602127	0.572383
Mining	−0.3933	−0.43774	0.134159	−0.43774
Manufacturing	−0.03346	−0.0579	0.180391	−0.0579
Exports indicator	0.332206	0.253317	0.373429	0.253317
Any intellectual property rights	0.91012369**	0.91129505**	0.327075	0.91129505**
Financial constraints	0.521218	0.51326	0.69744064**	0.51326
Source: Foreign subsidiary	−0.17302		0.334381	
Percent of total employees with higher education greater than 25 percent	−0.27162	−0.26049	0.019422	−0.26049
Nonsubsidiary		−0.46823		−0.46823
Constant	−.94264172**	−0.49194	−1.0731374**	−0.49194
N	914	914	905	914
Rank	11	12	11	12

Source: Author calculations, data from NIS database.
Note: Reference categories: small firm; service sector; domestic subsidiary.
Significance levels: * = 10 percent, ** = 5 percent, *** = 1 percent.

Table 2C.2 Introduction of a New Product and New Process, Innovators-Only Subsample

Variable	New product	New product	New process	New process
Micro	1.7291401**	1.7411317**	-0.51086	-0.41761
Medium	1.0345562**	1.0361838**	-0.17017	0.288265
Large	0.376814	0.370462	0.164767	0.48183
Mining	-0.53495	-0.57909	0.501904	0.467675
Manufacturing	0.147761	0.147531	0.449518	0.173881
Exports indicator	0.85218575**	0.87095395**	0.69550549**	0.68484933**
Any intellectual property rights	1.2821368***	1.2580294***	0.097368	-0.22635
Financial constraints	-0.15965	-0.19194	0.497529	0.68017388**
R&D	1.1558644***	1.1674568***	1.0558037***	1.1369636***
Foreign subsidiary	-0.99758914*	-0.99680831*	0.104602	0.409327
Nonsubsidiary	0.31446	0.306192	0.005623	0.257005
Percent of total employees with higher education greater than 25 percent	-0.61413571*	-0.64756351*	-0.4178	0.432754
Any public support	0.726341	0.724851	-0.86484	-1.1293775**
Cooperation on innovation	1.359606***	1.3554482***	0.623554	0.76949242**
Acquisition of machinery and equipment		-0.12143		1.6433741***
Acquisition of other external knowledge	0.069999			0.462977
Constant	-1.2107675*	-1.1107	-0.66654	-2.470185***
N	562	562	554	554
Rank	15	17	15	17

Source: Author calculations, data from NIS database.
Note: Reference categories: small firm; service sector; domestic subsidiary.
Significance levels: * = 10 percent, ** = 5 percent, *** = 1 percent.

Table 2C.3 Marketing Changes: Full Sample Probability Weights

Variable	Marketing change				
Micro	0.419211	0.419278	0.46085	0.260501	0.307636
Medium	0.7078306**	0.70903232**	0.65074088**	0.98601418***	0.96026095***
Large	1.1227891***	1.1184162***	1.0750124***	.59457467*	0.572383
Mining	-1.1022104***	-1.0886666**	-1.1732874**	-0.40798	-0.43774
Manufacturing	-0.21694	-0.22541	-0.2541	-0.04512	-0.0579
Exports indicator	0.64135202**	0.69148802**	0.69636214**	0.247776	0.253317
Any intellectual property rights	0.161177	0.190578	0.177573	0.92121483**	0.91129505**
Financial constraints	0.256802	0.249218	0.274183	0.489092	0.51326
Foreign subsidiary: indicator		-0.17745	-0.18054		
Percent of total employees with higher education greater than 25 percent			-0.37239		
Foreign subsidiary				-0.56059	-0.54537
Nonsubsidiary				-0.48723	-0.46823
Constant	-1.2026074***	-1.1956583***	-1.0926737***	-0.53137	-0.49194
N	936	928	915	927	914
Rank	9	10	11	11	12

Source: Author calculations, data from NIS database.

Note: Reference categories: small firm; service sector; domestic subsidiary.
Significance levels: * = 10 percent; ** = 5 percent; *** = 1 percent.

Table 2C.4 Organizational Changes: Full-Sample Probability Weights

Variable			Organizational change		
Micro	−0.68316	−0.69678	−.9956393**	0.250501	0.307636
Medium	0.261773	0.258988	0.467428	0.98601418***	.96026095***
Large	0.76115345**	0.79007759**	1.0049792**	0.59457467*	0.572383
Mining	−0.4199	−0.55847	−0.47371	−0.40798	−0.43774
Manufacturing	−0.43486	−0.41809	−0.38449	−0.04512	−0.0579
Exports indicator	0.59346242*	0.395545	0.337844	0.247776	0.253317
Any intellectual property rights	0.103254	0.009846	0.038751	0.92121483**	0.91129505**
Financial constraints	1.0893513***	1.1302222***	1.1600489***	0.489092	0.51326
Foreign subsidiary: indicator	0.80874209***	0.92003296**			
Percent of total employees with higher education greater than 25 percent			0.85268228**		−0.026049
Foreign subsidiary				−0.56059	−0.54537
Nonsubsidiary				−0.48723	−0.46823
Constant	−0.3156	−0.34005	−0.60948	−0.53137	−0.49194
N	850	843	831	927	914
Rank	9	10	11	11	12

Source: Author calculations, data from NIS database.

Note: Reference categories: small firm; service sector; enterprise part of a larger group: domestic head office.

Significance levels: * = 10 percent, ** = 5 percent, *** = 1 percent.

Table 2C.5 Marketing Innovation: Innovator-Only Subsample Probability Weights

Variable	Marketing change			
Micro	1.2562541*	1.3531392**	1.554606**	1.8295234***
Medium	1.0420869**	1.1533718***	1.081092**	1.2663495***
Large	1.0926058**	1.0632517**	1.041284**	1.091628**
Mining	−1.3605714**	−1.4788647**	−1.457933*	−1.3286425*
Manufacturing	−0.44196	−0.69077775*	−0.58613951*	−0.61909035*
Exports indicator	0.615894	0.55647184*	0.546337	0.56525025*
Any intellectual property rights	−0.17357	−0.14592	−0.11583	−0.12074
Financial constraints	−0.34455	−0.38397	−0.22581	−0.22645
R&D	0.455139	0.505038	0.527299	.60082164*
Foreign subsidiary		−1.164217*	−1.02222	−0.9686
Nonsubsidiary		−1.0389163**	−0.88343705**	−0.78754096**
Percent of total employees with higher education greater than 25 percent			−0.8485833*	−0.66022
Any public support			−0.16847	−0.24435
Cooperation on innovation			0.048115	0.142132
Acquisition of machinery and equipment				0.75331061*
Acquisition of other external knowledge				−0.07005
Constant	−.95283017*	0.016673	−0.11933	−0.99864
N	583	575	563	563
Rank	10	12	15	17

Source: Author calculations, data from NIS database.

Note: Reference categories: small firm; service sector; domestic subsidiary.

Significance levels: * = 10 percent, ** = 5 percent, *** = 1 percent.

Table 2C.6 Organizational Innovation: Innovator-Only Subsample Probability Weights

Variable	Organizational change			
Micro	−1.23447	−2.3831365***	−2.5043875***	−2.4172702***
Medium	0.375059	−0.19248	0.062707	0.249065
Large	−0.00438	−0.22777	−0.04479	0.066143
Mining	1.5344047**	1.110519	1.282767	1.690214
Manufacturing	−0.62908	−0.43669	−0.38593	−0.54706
Exports indicator	0.683693	0.180666	0.159702	0.172056
Any intellectual property rights	−0.69902994*	−0.9233139**	−1.0330775**	−0.09244073**
Financial constraints	1.4671073***	1.3260556***	1.0244707***	1.2119347***
R&D	1.2043818***	1.0394159***	0.97929306***	0.95316054***
Foreign subsidiary		1.7667958***	1.6782497***	1.7706643***
Nonsubsidiary		0.333711	0.252452	0.293436
Any public support		−1.8046462***	−1.5223756***	−1.7044848***
Cooperation on innovation		1.6054983***	1.4928656***	1.541504***
Percent of total employees with higher education greater than 25 percent			0.89510853**	1.2329003**
Acquisition of machinery and equipment				0.69283377*
Acquisition of other external knowledge				0.119175
Constant	0.019184	0.122552	0.024776	−0.66464
N	528	518	510	510
Rank	10	14	15	17

Source: Author calculations, data from NIS database.

Note: Reference categories: small firm; service sector; domestic subsidiary.

Significance levels: * = 10 percent, ** = 5 percent, *** = 1 percent.

Table 2C.7   New to Market versus New to Firm

Variable	New to market	
Micro	−1.2272428**	−1.3067231**
Medium	−0.46834	−0.44022
Large	0.533976	0.602271
Mining	−2.1156343***	−2.142864***
Manufacturing	0.72370855**	0.72050711**
Exports indicator	−1.272434***	−1.3425291***
Any intellectual property rights	0.221625	
R&D	1.5291298***	1.4493825***
Financial constraints	0.056373	0.116984
Any public support	−1.0127186*	−1.023995*
Cooperation on innovation	0.062725	−0.06894
Foreign subsidiary	1.4302072***	1.3185767**
Nonsubsidiary	0.555048	0.65255413*
Percent of total employees with higher education greater than 25 percent	0.94024079***	1.0049378***
Acquisition of machinery and equipment	−0.36604	−0.46205
Acquisition of other external knowledge	0.95570684***	0.87356169**
Patent in South Africa		−0.29954
Patent outside South Africa		0.841281
Registered design		1.1524352**
Registered trademark		−0.09281
Claimed copyright		0.605545
Grant license on IPR resulting from innovation		−0.15842
Constant	−0.77772	−0.66389
N	458	458

Source: Author calculations, data from NIS database.

Note: Reference categories: small firm; service sector; domestic subsidiary. The dependent variable here is one if a firm introduces new-to-market innovations and zero if it introduces new-to-firm innovations. Significance levels: * = 10 percent, ** = 5 percent, *** = 1 percent.

Table 2C.8   Determinants of Research and Development

Variable	R&D	
Micro	−0.97945011*	−0.9664336*
Medium	−0.13485	−0.05611
Large	−0.14147	−0.05591
Mining	0.414375	0.483896
Manufacturing	0.81376084**	0.77076975**
Exports indicator	0.92233779***	0.94044767***
Any intellectual property rights	−0.52255	
Financial constraints	−0.3867	−0.43573
Foreign subsidiary	−0.90609	−1.2299492**
Nonsubsidiary	−0.66682	−0.57787
Any public support	0.688415	0.748631
Cooperation on innovation	1.6893893***	1.688476***
Percent of total employees with higher education greater than 25 percent	−0.18744	−0.31584
Acquisition of machinery and equipment	−0.08418	−0.26023
Acquisition of other external knowledge	0.69771789*	0.619372
Patent in South Africa		1.5748689***
Patent outside South Africa		0.366269
Registered design		2.2504201**
Registered trademark		−1.3625215***
Claim copyright		0.496461
Grant license on IPR resulting from innovation		−0.01001
Constant	0.335776	0.438276
N	563	563
Rank	16	21

Source: Author calculations, data from NIS database.
Note: Reference categories: small firm; service sector; domestic subsidiary.
Significance levels: * = 10 percent, ** = 5 percent, *** = 1 percent.

Table 2C.9   Factors Hampering Innovation Activities, 2002–04, Weighted Percentages

Constraints to innovation activities	Not relevant	Low	Medium	High	Missing	Total
Lack internal funds	43	12	17	25	3	100
Lack external funds	54	15	13	15	3	100
Innovation costs too high	45	13	19	20	3	100
Lack qualified personnel	46	13	21	17	3	100
Lack of information on technology	50	16	27	4	3	100
Lack of information on markets	53	16	24	4	3	100
Difficulty in finding co-op partner	58	18	14	8	3	100
Market dominated by established firms	47	9	15	26	3	100
Uncertain demand for innovative goods or services	48	16	24	10	3	100
No need because of prior innovations	65	19	10	4	3	100
No need because no demand for innovations	61	16	11	9	3	100

Source: Author calculations, data from NIS database.

Table 2C.10 Determinants of Technology Absorption: South Africa, ES 2003 Data

Variable	Product innovation	Process innovation	Product innovation	Process innovation	Product innovation	Process innovation
Firm age	−0.001	0.001	−0.002	0.002	−0.003	0.001
Firm – medium	0.353	0.052	0.297	−0.005	0.287	−0.017
Firm – large	0.537**	0.296	0.470*	0.197	0.461*	0.164
Majority foreign firm	−0.233	−0.150	−0.267	−0.223	−0.307	−0.254
Domestic multinational	−0.093	−0.033	0.131	−0.004	0.120	−0.025
ISO certification	0.324**	0.122	0.321**	0.066	0.262*	0.057
Access to credit	0.044	0.233*	0.140	0.346***	0.133	0.330**
Business association	0.036	0.121	−0.045	0.052	−0.070	0.046
R&D dummy	0.688***	0.426***	0.654***	0.338***	0.630***	0.326**
Web and e-mail	0.127	0.889	0.056	0.990	−0.007	0.977
High-tech sector	0.216	−0.023	0.222	−0.034	0.206	−0.057
Competition – Weak	−0.512	−0.616*	−0.424	−0.545	−0.434	−0.542
Competition – Medium	−0.383	−0.547	−0.250	−0.512	−0.221	−0.516
Competition – Strong	0.046	−0.498	0.170	−0.479	0.214	−0.485
Manager with tertiary education			0.397**	0.543***	0.388**	0.539***

Manager's years of experience	−0.050		0.015**	0.006	0.015**	0.005
Worker training provided	−0.179		−0.010	0.181	−0.017	0.176
Share of skilled employees	−0.056		0.003	−0.003	0.003	−0.003
Import dummy					0.124	0.166
Licensed foreign technology					0.126	0.053
Joint venture with foreign company					0.477*	0.031
Export dummy (2000, 2001, 2002)		−0.033	−0.086	−0.042	−0.137	−0.063
Non-African exports		0.015	−0.177	−0.044	−0.156	−0.048
African exports		0.121	−0.063	0.065	−0.089	0.044
Durban	0.127	0.278	0.162	0.306	0.128	0.309
Johannesburg	0.193	0.243	0.289	0.229	0.318*	0.248
Port Elizabeth	0.152	0.209	0.006	0.128	0.013	0.158
Constant	−0.403	−0.915	−1.002	−1.389*	−0.990	−1.420*
N	508	508	493	493	493	493

Source: Author calculations, ES 2003 data.

Note: Likelihood ratio test of rho = 0 always has *p*-value < 0.005. Reference categories: small firm; competition – none; Cape Town.

Significance levels: * = 10 percent, ** = 5 percent, *** = 1 percent.

Table 2C.11 Determinants of Technology Absorption: South Africa, ES 2007 Data

Variable	Product innovation	Process innovation	Product innovation	Process innovation	Product innovation	Process innovation
Firm age	0.004	0.008*	0.002	0.007	0.002	0.008
Firm – medium	0.447**	0.396**	0.362*	0.353*	0.302	0.310
Firm – large	0.365	0.162	0.183	0.078	0.116	0.050
Majority foreign firm	0.562**	0.027	0.533**	0.041	0.525**	0.038
Intermediate sales	−0.003	0.000	−0.003	0.000	−0.005	−0.001
ISO certification	0.382**	0.038	0.363**	0.044	0.423**	0.095
Access to credit	0.113	0.063	0.043	−0.004	0.028	−0.040
Web	0.039	0.158	−0.046	0.163	0.002	0.197
High-technology sector	−0.172	−0.180	−0.174	−0.185	−0.117	−0.156
Low competition	−0.030	0.116	−0.044	0.108	0.009	0.145
Medium-high competition	0.207	0.303	0.183	0.257	0.204	0.302
Competition with informal firms	0.583***	0.168	0.577***	0.156	0.631***	0.162
Manager with tertiary education			0.200	−0.188	0.217	−0.177
Manager's years of experience			0.007	0.011	0.011	0.013*

Worker training provided		0.294*	0.212	0.248	0.189
Share of skilled employees		−0.001	−0.002	−0.001	−0.002
Percentage exports (2006)				0.004	0.015***
Import dummy				0.090	0.015
Licensed foreign technology				0.151	−0.074
Firm located in export/industrial zone				−0.411***	−0.287*
Dummy for exports to SADC	−0.005	−0.014	−0.074	−0.133	−0.284
Exports to developed countries	−0.066	−0.130	−0.143	−0.212	−0.365
Durban	−0.687**	−0.596**	−0.346	−0.566**	−0.243
Johannesburg	0.119	0.154	0.386*	0.101	0.432*
Port Elizabeth	−0.897***	−0.797**	−0.337	−0.802***	−0.364
Constant	−1.660***	−1.256***	−1.632***	−1.188***	−1.678***
N	420	419	419	419	419

Source: Author calculations, ES 2007 data.

Note: Likelihood ratio test of rho=0 always has *p*-value <0.001. Reference categories: small firm; competition – none; Cape Town.

Significance levels: * = 10 percent, ** = 5 percent, *** = 1 percent.

Table 2C.12 Probability of Introducing a New Product in the Past Three Years, Mauritius, ES 2008 Data

Variable	Introduction of a new product			
Small	−0.157	0.07		
	[0.086]*	[0.112]		
Medium	−0.066	0.098	0.186	0.186
	[0.078]	[0.101]	[0.085]**	[0.085]**
Large			0.091	0.091
			[0.112]	[0.112]
Age	0.002	0.002	0.001	0.001
	[0.001]	[0.002]	[0.002]	[0.002]
ISO	0.187	0.068		
	[0.092]**	[0.139]		
R&D	0.371	0.357	0.327	0.327
	[0.061]***	[0.081]***	[0.084]***	[0.084]***
Training		0.241	0.218	0.218
		[0.084]***	[0.083]***	[0.083]***
Exporter			−0.119	−0.121
			[0.092]	[0.096]
Family owned			−0.011	−0.011
			[0.123]	[0.123]
Foreign ownership				0.009
				[0.162]
Observations	321	221	226	226

Source: Author calculations, ES 2008 data.
Note: Standard errors are in brackets. All regressions include sector dummies.
Significance levels: * = 10 percent, ** = 5 percent, *** = 1 percent.

Table 2C.13 Probability of Introducing a New Technology in the Past Three Years, Mauritius, ES 2008 Data

Variable	Introduction of a new technology				
Small	−0.156	−0.157	−0.159	−0.251	−0.254
	[0.072]**	[0.072]**	[0.073]**	[0.086]***	[0.086]***
Large	0.094	0.104	0.112	0.025	0.029
	[0.081]	[0.082]	[0.082]	[0.107]	[0.108]
Age	0.001	0.001	0.001	0.001	0.001
	[0.001]	[0.001]	[0.001]	[0.002]	[0.002]
R&D	0.426	0.431	0.437	0.452	0.451
	[0.068]***	[0.068]***	[0.068]***	[0.093]***	[0.093]***
ISO	0.2	0.215	0.223	0.296	0.298
	[0.104]*	[0.105]**	[0.106]**	[0.156]*	[0.156]*
Exporter		−0.068	−0.042	−0.124	−0.121
		[0.083]	[0.090]	[0.109]	[0.109]

(continued next page)

Table 2C.13   *(continued)*

Variable	Introduction of a new technology				
Foreign ownership			−0.092	−0.154	−0.154
			[0.109]	[0.144]	[0.144]
Training				0.246	0.25
				[0.096]**	[0.096]***
Family owned					0.07
					[0.134]
Observations	306	304	304	214	214

Source: Author calculations, ES 2008 data.
Note: Standard errors are in brackets. All regressions include sector dummies.
Significance levels: * = 10 percent, ** = 5 percent, *** = 1 percent.

Notes

1. He finds evidence of a "feedback mechanism" from exporting to productivity. Exporters pay higher wages, operate on a larger scale, are more capital intensive, and have higher productivity. He also shows that the new exporters increase the productivity gap relative to nonexporters once they enter foreign markets.

2. The Industrial Policy Action Plan (IPAP) summarized the situation as follows: "A relatively small number of automotive components dominate the export basket and local content has stagnated" (Government of South Africa 2010a, 55).

3. The nominal effective exchange rate of the rand with 15 trading partners was 90.84 in 2004 and 64.70 in 2008 (South African Reserve Bank data).

4. The SAABC has 63 firms on its database. Of those, 23 firms declared that they had a budget for R&D. The total number of local auto component firms is in the region of 200.

5. Cost per employee for South Africa compared with Thailand was as follows: artisans (12 times); professionals (6 times); supervisors (4 times); apprentices (almost 3 times); and management (almost 3 times) (Durban Automotive Cluster 2011).

6. In the ES, more nonexporters undertake training than exporters. This finding is expected, because nonexporters ranked the shortage of skills as their key constraint. Of exporters, 69 percent trained for an average of 41 days for a skilled worker and 44 days for an unskilled worker; of nonexporters, 86 percent trained for an average of 34 days for a skilled worker and 47 days for an unskilled worker.

7. These data are purely for the equipment exports. The South African Revenue Service does not monitor the export sales figures for services, so those figures

are not included (information from South African Capital Equipment Export Council Business Plan 2009–10 provided by interview respondent).

8. According to the IPAP, the primary chemical sector accounted for 3.2 percent of South Africa's GDP (excluding plastics and rubber) (Government of South Africa 2010a, 60–61).

9. Data are from the Chemical and Allied Industries' Association (CAIA) website, http://www.caia.co.za/.

10. The average fixed broadband Internet rate is US$26.30 per month, compared with US$6 per month in India and US$19 per month in China. International bandwidth averages 71 bits per second.

11. The financial service provider was initially a domestic firm specializing in electronic payment systems, used in 31 countries. It was bought out by a NASDAQ-listed U.S. financial service provider in 2004.

12. The Black Empowerment Charter for the IT sector requires firms to comply with the Employment Equity Act, specifically to prioritize strategic positions or jobs for black candidates. These requirements include targets of 50 percent black people in senior management positions and 65 percent black people in other management positions (30 percent of these should be black women). The charter also requires that firms provide for skill development, training, and mentoring, and that 30 percent to 70 percent of procurement spending be directed toward black-owned, black-empowered suppliers (ICT Charter Steering Committee 2005).

13. R&D and new–to-the-market versus new-to-the-firm regressions are run in the subsample.

14. The Likert scale is a methodology to determine attitudes along a continuum of choices, such as "strongly agree," "agree," and "strongly disagree."

15. Table 2C.2 in annex 2C also differentiates firms that introduced product innovations from those that did not, because even successful innovators also reported these constraints.

16. Aerts and Schmidt 2008; Almus and Czarnitzki 2003; Bérubé and Mohnen 2009; Busom 2000; Czarnitzki, Ebersberger, and Fier 2006; Czarnitzki and Licht 2006; González, Jaumandreu, and Pazó 2005; Hall and Maffioli 2008— all cited in Mairesse and Mohnen 2010.

17. Heineken and Diageo have created a 50:50 joint venture to hold a 28.9 percent stake in the brewery.

18. Some fish-processing firms in Namibia reportedly engage only in commodity products using much less sophisticated processing technology. These firms, purportedly foreign owned, then undertake more sophisticated processing abroad. This case study did not include interviews with any such firms.

19. These 16 firms had previously been surveyed in the 2008 ES and had claimed to introduce a new product or process.

20. The Taiwanese and South African firms in Lesotho had internal rework rates of 10.83 percent and 8.10 percent, respectively, much higher than that of firms in neighboring South Africa (6.31 percent). The figures for customer returns were 0.47 percent for Taiwanese firms, 1.86 percent for South African firms in Lesotho, and 0.63 percent for South African firms.

21. The question was, "During the 3-year period 2002–2004, did this business introduce new or significantly improved goods?" and similarly for the others (see Blankley and Moses 2009, Appendix 5: Questionnaire).

References

AIEC (Automotive Industry Export Council). 2009. http://www.aiec.co.za/.

Almeida, Rita, and Ana Margarida Fernandes. 2008. "Openness and Technological Innovations in Developing Countries: Evidence from Firm-Level Surveys." *Journal of Development Studies* 44 (5): 701–27.

Blankley, William, and Cheryl Moses. 2009. *Main Results of the South African Innovation Survey 2005.* Cape Town: Human Sciences Research Council Press.

Buys, Andre. 2010. "Ownership and Innovative Behaviour. The Case of the South African Automotive Component Manufacturing Industry." In *Technology Management for Global Economic Growth (PICMET) 2010 Proceedings,* Phuket, Thailand, July 18–22.

Crépon, Bruno, Emmanuel Duguet, and Jacques Mairesse. 1998. "Research, Innovation and Productivity: An Econometric Analysis at the Firm Level." *Economics of Innovation and New Technology* 7 (2): 115–58.

Durban Automotive Cluster. 2011. http://www.dbnautocluster.org.za/dac/tmp/The%20future%20of%20the%20SA%20automotive%20industry%20under%20APDP%20-%20Dr%20Justin%20Barnes.pdf.

Enterprise Surveys (database). World Bank, Washington, DC. https://www.enterprisesurveys.org.

Garcia, Abraham, and Pierre Mohnen. 2010. "Impact of Government Support on R&D and Innovation." UNU-MERIT Working Paper 2010-034, United Nations University, Maastricht, The Netherlands.

Gastrow, Michael. 2008. "An Overview of Research and Development Activities in the South African Automotive Industry." *Journal for New Generation Sciences* 6 (1): 1–15.

Goldberg, Itzhak, Lee Branstetter, John G. Goddard, and Smita Kuriakose. 2008. "Globalization and Technology Absorption in Europe and Central Asia: The Role of Trade, FDI, and Cross-Border Knowledge Flows." Working Paper 150, World Bank, Washington, DC.

Government of Mauritius. 2010. "Industrial and SME Strategic Plan 2010–2013." Ministry of Industry and Cooperatives, Port-Louis, Mauritius.

Government of South Africa. 2010a. "2010/11–2012/13 Industrial Policy Action Plan 2: February 2010." Department of Trade and Industry, South Africa.

———. 2010b. Sector database. http://www.dti.gov.za.

Griffith, Rachel, Stephen Redding, and John Van Reenen. 2004. "Mapping the Two Faces of R&D: Productivity Growth in a Panel of OECD Industries." *Review of Economics and Statistics* 86 (4): 883–95.

Habyarimana, James, Giuseppe Iarossi, Michael Ingram, Laura Klapper, Ganesh Rasagam, Anne Rennie, and Manju Shah. 2005. *Mauritius: Investment Climate Assessment.* Washington, DC: World Bank.

ICT Charter Steering Committee. 2005. "Draft Black Economic Empowerment Charter for the ICT Sector." Final Version May 2005. Pretoria, South Africa. http://www.ictcharter.org.za/.

Kadhikwa, Gerson, and Vitalis Ndalikokule. 2007. "Assessing the Potential of the Manufacturing Sector in Namibia." Bank of Namibia Occasional Paper 1/2007, Windhoek, Namibia.

Kuriakose, Smita, Itzhak Goldberg, and David Kaplan. 2009. "Innovation and Technology Absorption." In *Mauritius: Investment Climate Assessment*, 51–77. Report No. 52794. Washington, DC: World Bank.

Mairesse, Jacques, and Pierre Mohnen. 2005. "The Importance of R&D for Innovation: A Reassessment Using French Data." NBER Working Paper 10897, National Bureau of Ecnomic Reasearch, Cambridge, MA. http:// www.nber.org/papers/w10897.

———. 2010. "Using Innovations Surveys for Econometric Analysis." NBER Working Paper 15857, National Bureau of Economic Research, Cambridge, MA.

MAIT (Manufacturers' Association for Information Technology). 2009. "ICT Market in South Africa." *MAIT Country Intelligence Report*, Vol. 101 (February).

Mashego, Tshepo. 2010. "Broadband Infraco Feels Pressure as Bandwidth Prices Fall Sharply." *BusinessDay*, allAfrica.com, August 20.

Mohnen, Pierre, Jacques Mairesse, and Marcel Dagenais. 2006. "Innovativity: A Comparison across Seven European Countries." UNU-MERIT Working Paper Series #2006-027, United Nations University, Maastricht, The Netherlands.

Morris, Mike, Cornelia Staritz, and Justin Barnes. 2011. "Value Chain Dynamics, Local Embeddedness and Upgrading in the Clothing Sectors of Lesotho and Swaziland." Draft for publication forthcoming in *International Journal of Technological Learning, Innovation and Development*.

NAACAM (National Association of Automotive Component and Allied Manufacturers). 2010. Industry Statistics. http://naacamdirectory.webhouse .co.za/pages/32981.

OECD (Organisation for Economic Co-operation and Development). 2010. "Ministerial Report on OECD Innovation Strategy." OECD Innovation Strategy, OECD, Paris.

Stijger, Roel M., and Jasper L. Steyn. 2010. "Contribution of Support Schemes to Innovativeness in the South African Automotive Component Industry." In *Technology Management for Global Economic Growth (PICMET) 2010 Proceedings*, Phuket, Thailand, July 18–22.

Van Biesebroeck, Johannes. 2005. "Exporting Raises Productivity in Sub-Saharan African Manufacturing Firms." *Journal of International Economics* 67 (2): 373–91.

Van Zyl, W. H., and Bernard Prior. 2009. "South Africa's Biofuels: IEA Taskgroup 39 Progress Report." Department of Microbiology, Stellenbosch University, South Africa.

World Bank. 2010a. *Mauritius: Enhancing and Sustaining Competitiveness: Policy Notes on Trade and Labor*. Report No: 53322-MU. Washington, DC: World Bank, Poverty Reduction and Economic Management, Africa Region.

———. 2010b. South Africa IT Sector Fact Sheet. World Bank, Washington, DC.

———. 2010c. *World Development Indicators*. Washington, DC: World Bank.

———. 2011a. "Closing the Skills and Technology Gaps in South Africa" (Draft May). World Bank, Washington, DC.

———. 2011b. "Skills and Technology Absorption in Mauritius" (Draft May). World Bank, Washington, DC.

———. 2011c. *Doing Business 2011: Making a Difference for Entrepreneurs*. Washington, DC: World Bank.

CHAPTER 3

Policy Options for Greater Technology Absorption

Chunlin Zhang, Itzhak Goldberg, David Kaplan, and Smita Kuriakose

This book has focused on technology absorption in a number of southern African countries that obtain their technology and ideas from abroad, principally from the industrialized countries. As was clear in chapter 2, although firms in these countries are generally able to identify and gain access to the requisite technologies, they are all faced with constraints that hinder their ability to effectively absorb technology. This chapter accordingly assesses policy options that are open to the governments and the private sector in southern African countries to address these constraints to improve technology absorption among firms.

Governments can support technology absorption in a variety of ways (see box 3.1 for a brief history of innovation policies among countries in the Organisation for Economic Co-operation and Development [OECD]). At the most basic level, effective government policies could create an institutional base that establishes openness to trade, improves the business environment for domestic and foreign investment, establishes effective intellectual property rights regimes, and enhances knowledge flows and learning. Beyond those general policies, many governments have also intervened at the industry and firm levels to address market failures.

Box 3.1

A Brief History of Innovation Policy in OECD Countries

In the first part of the 20th century, innovation policy as such did not truly exist. It was gradually developed as a way to promote the industrial competitiveness and social welfare of countries, as a complement to actions taken by governments to develop defense technologies, as initiated in World War II. Innovation policy has emerged gradually as a policy distinct from both science and industry policies. The evolution of government efforts to encourage innovation over the second part of the 20th century can be summarized as follows.

1950s. This decade saw the building of modern science systems in the industrialized world. In some countries, piecemeal measures were occasionally adopted to reduce identified weaknesses in the innovation process, including the creation of the National Research and Development Corporation in the United Kingdom (1949), the aim of which was to facilitate the promotion and diffusion of inventions from public laboratories and universities. Among others, France established sector-specific technical centers to help industries with technical research assistance and information, and Germany set up the Fraunhofer system of applied research and development (R&D).

1960s. Two trends were noticeable. First was the launch of large-scale programs in strategic areas such as space, nuclear technology, and oceanography, in countries such as France, the United Kingdom, and the United States. Second was the emergence of the concept of innovation policy distinct from science policy. The seminal report in this area is the Charpie Report published in the United States at the request of the Department of Commerce in 1967 (U.S. Department of Commerce 1967). It stated clearly the need to act on diverse factors affecting the innovation climate, such as university-industry relations, venture capital, procurement policies, tax incentives, and competition laws—with particular attention to small enterprises and individual inventors, which it presented as the main source of innovation.

1970s. The decade saw a proliferation of government measures to promote innovation in the form of civilian technology programs, R&D incentive schemes for in-house efforts in the business sector, and university-industry collaboration. This was particularly evident in Europe and Japan, which were trying to decrease the increasing technology gap with the United States. The oil crisis of 1973–75, and the subsequent economic slowdown, also led to renewed interest in innovation policies. A concerted effort was made to develop an institutional

(continued next page)

Box 3.1 *(continued)*

framework for innovation. For instance, in the United States, the Small Business Administration was strongly involved in support for small firms, which were perceived as a key source of innovation (as in the Charpie Report); the National Science Foundation, for its part, supported basic research; and the various sector agencies (defense, commerce, interior, and so on) all had technology-related programs.

1980s. Two major trends emerged. First was the development of regional technology and innovation policies, owing to an increased perception that innovation flourishes in sites with a concentration of talent, knowledge, and resources. Therefore, building critical mass was considered important, and major programs were set up to build science parks or "technopolises" (in Japan, for example). The need was also felt to act as closely as possible to entrepreneurs and potential innovators in order to help them more efficiently. Hence, territorially decentralized innovation policy initiatives proliferated, often encouraged by central governments through various schemes (such as decentralized antennas of central innovation agencies, or matching funds provided to local governments). The second major feature of this decade was the emergence of the notion of national innovation systems, which emphasized the interactions among key actors and communities (research, business, education) as a source of the innovative dynamism of countries and the need for governments to strengthen such systems through appropriate policy actions.

1990s. Inspired by the concept of national innovation systems as well as the acceleration of the globalization process, the spread of information and telecommunication technologies, and the emergence of new technologies such as biotechnologies, governments systematically engaged in building innovation policies that encompassed established policy fields. In the traditional science policy field, efforts were made to connect basic research more closely to applications. In industry policy, horizontal actions to boost innovation efforts were perceived as an efficient way to replace traditional policies to "pick winners," which were criticized for their inefficiency and ideological inadequacy. The Nordic countries have probably been the most active in adopting this approach. In the mid-1990s, Finland, for example, created two key institutions for promoting innovation: Tekes, the technology agency in charge of supporting innovation directly with a very significant budget; and the Science and Technology Policy Council, chaired by the prime minister, with the active participation of all ministers (including finance),

(continued next page)

Following on the findings presented in chapter 2, this chapter further explores public policy options. First, the government needs to get the basics right. Second, more-focused policy actions can be taken in four particular areas: learning through trade, foreign direct investment (FDI) spillovers, skill development, and research and development (R&D) activities.

Getting the Basics Right

This book has shown that constraints to technology absorption in southern African firms do not differ very much from those to economic growth in general. Before taking more proactive actions to correct market failures, governments in the four studied countries are well advised to get the basics right.

Nurturing Entrepreneurship

Governments play an important role in creating an environment for entrepreneurs conducive to creating wealth through innovation by removing bottlenecks in the general business environment that impede those able entrepreneurs with good ideas from starting a new venture and creating jobs. Such support can include a business environment that allows failure and company exit as a necessary part of entrepreneurial learning, company incentives that favor entrepreneurs with good ideas, instruments that enable entrepreneurs to access capital for start-ups, and flexible labor market policies that enable firms to expand by attracting the most skilled and talented personnel from outside the firm or the country.

Strengthening Competition

Greater competition is particularly critical for technology innovation and absorption in South Africa. In the National Innovation Survey (NIS), of the constraints listed, the greatest number of firms (26 percent) cited the market being dominated by established firms as a reason for not innovating. In one of the firms interviewed in the study, the management indicated clearly that the competitiveness of its products depends not so much on innovation and technology upgrading as it does on the price of one input, which accounts for 75 percent of the total cost of the product and is supplied by one dominating domestic firm. This observation adds to one of the key findings of the recently completed second Investment Climate Assessment of South Africa, namely, the need for a more activist and innovative competition policy aimed at tackling barriers to entry and innovation. A widely shared view is that industrial concentration in South Africa has hindered market competition, with adverse implications for productivity and employment in all sectors (for example, see Government of South Africa [2010]). In addition to competition policy, trade and investment policies should be used to generate positive change in the market structure and to introduce competition pressure from abroad.

Improving Investment Climate

A concerted government effort is also needed to improve the investment climate in southern African countries. A vast majority of the firms interviewed were constrained by logistical and infrastructural factors. High costs and lack of reliable rail transportation, harbor access, and power provision skew technological choices by favoring road transport or the installation of company-owned power generators, for example, adversely affecting exports. Although many South African firms are efficient and use technology well in house, they face key constraints in the broader environment within which they operate. Another example is the pricing of energy. Subsidized energy prices discourage energy-saving technology progress. Therefore, getting the energy prices right is essential to give firms incentives to invest in energy-saving green technologies. The forthcoming Industrial Competitiveness and Upgrading Program in South Africa will provide incentives for firms to invest in green and energy-saving technologies.

Improvement in investment climate is particularly important for technology absorption through FDI. By strengthening the attractiveness of the host economy, a better investment climate enables the host country to negotiate more easily with foreign investors to increase technology and skill transfer.

Increasing Skill Supply

All the case studies in chapter 2 point to skill shortages as the leading constraint to greater technology absorption in manufacturing enterprises. This limitation is particularly severe in South Africa. Corroborating the study findings are employer perception surveys that relate to the availability of skills in South Africa and have similar findings.[1] Furthermore, the recent World Bank draft study on closing the skills and technology gaps (World Bank 2011a) has found evidence of skill gaps in the labor market to be very strong. Drawing on this parallel study[2] and using the framework presented in chapter 1, this section explores the causes of skill shortages in South Africa and discusses the direction of reform. This exploration is supplemented by a brief review of policy options in other countries.

Causes of Skill Shortages

One obvious cause of South Africa's skill shortages is its historical legacy. The previous apartheid regime represented an institutionalized program of educational inequality and separation based on race (World Bank 2011a). Under the apartheid system of "native education," black children were taught a different curriculum, with a view that they had no place "above the level of certain forms of labor."[3] The damage done by apartheid-era education has long-lasting effects, because skill development is a highly path-dependent process. The second cause, which is less frequently cited, is the rising skill intensity of labor demand. *Skill shortage* is, of course, a gap between supply of and demand for skilled labor and therefore relative. Indeed, the shortage of skill supply can also be seen as a problem of excessive demand: despite the shortage of skills, South Africa's economic growth since the mid-1990s has generated a labor demand with increasing skill intensity. According to a study of labor market survey data done by Bhorat and Mayet (2011), the number of degree-holding workers doubled during 1995–2009, while the number of workers with no education declined by 43.5 percent. In between, increases of 72 percent to 79 percent in employment were seen in those who completed grade 12, and a decline of 17 percent for those who did not complete General Education and Training.[4]

However, the main cause of the observed skill shortage in South Africa seems to be the inadequacy of supply-side response to demand. This shortage is evidenced by the failure of the technical and vocational education and training (TVET) system as well the higher education system in increasing the supply of skilled labor to keep pace with evolving demand.

First, overall capacity of the postsecondary education system, which is dominated by public higher education institutions (HEIs) and public further education and training institutions (FETIs), is extremely limited in relation to demand. According to an analysis of the data collected in the 2007 Community Survey (Sheppard and Cloete 2009), 2.8 million youth in the 18–24 year age cohort in 2007 were not in employment, education, or training (known as "NEETs"). In contrast, the total enrollment of public HEIs and FETIs in 2007 was only 1.1 million. In addition to further education and training and higher education, apprenticeship and artisan-related learnership are the third route for postschool youth to obtain education and training. However, enrollments in apprenticeships and learnerships, after some increase from previous years, reached only 0.03 million individuals in 2009–10, which is insignificant to the scale of the challenge. The capacity of the system would have to increase by 1.5 times if the 2.8 million NEETs were put in education or training in public institutions.

Second, the growth of capacity and output of the postsecondary education system does not seem to have responded well to the persistent skill shortage. Figure 3.1 provides a series of indicators in this regard. Panel a highlights the absence of growth in terms of teaching capacities in both public HEIs and public FETIs over the past decade. Panel b indicates an increase in public HEI enrollment of only 0.23 million in the decade. The total number of learners in public FETIs also increased in the same period, but again only by a very small margin of 0.15 million. Enrollments in the fields of business and management in public HEIs have not increased for a decade (panel d), nor has the enrollment of master's degree students (panel e). Although the number of enrollments in the fields of science, engineering, and technology has increased by 75,702 from 2000 to 2009 (panel c), the proportional headcount enrollment in universities of technology has declined since 2004. While enrollment in doctoral programs has increased modestly between 2005 and 2009, the number of PhD graduates has remained stagnant (panel f). All these increases pale in comparison with the scale of new labor market entrants of 1 million every year.

The lack of increase in enrollment is, however, compounded by strikingly low completion rates, as highlighted by the gaps between enrollment and graduation levels in figure 3.1. This general picture is well supported by more detailed studies. For example, of the national diploma students who entered in 2000, only about 15 percent graduated

Figure 3.1   Capacity and Output of the Postsecondary Public Education System, South Africa, 1999–2009

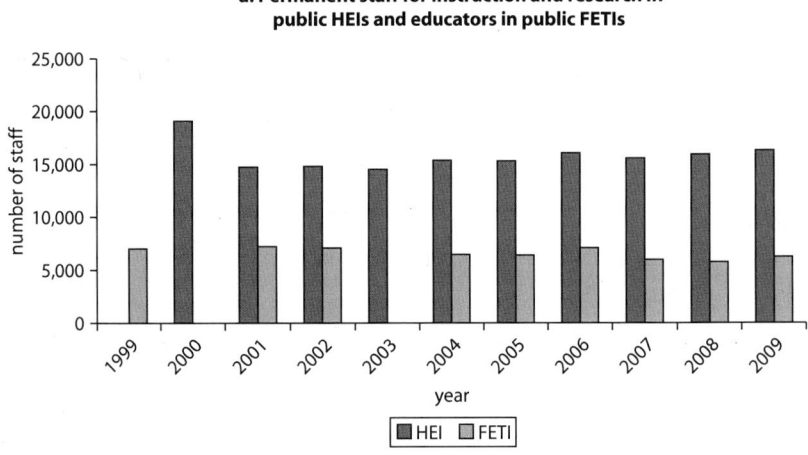

a. Permanent staff for instruction and research in public HEIs and educators in public FETIs

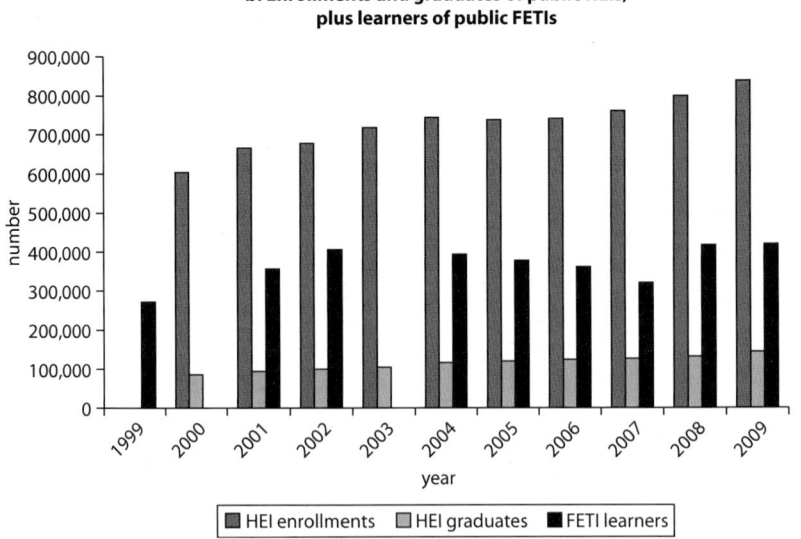

b. Enrollments and graduates of public HEIs, plus learners of public FETIs

(continued next page)

Figure 3.1   *(continued)*

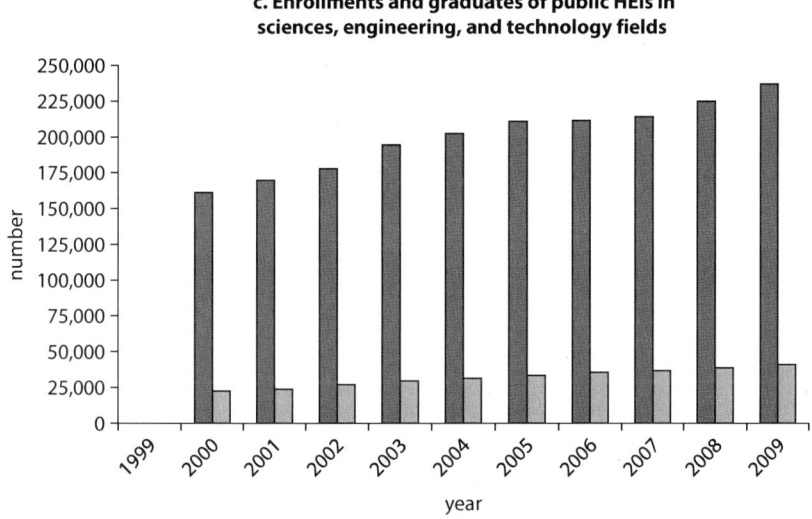

c. Enrollments and graduates of public HEIs in sciences, engineering, and technology fields

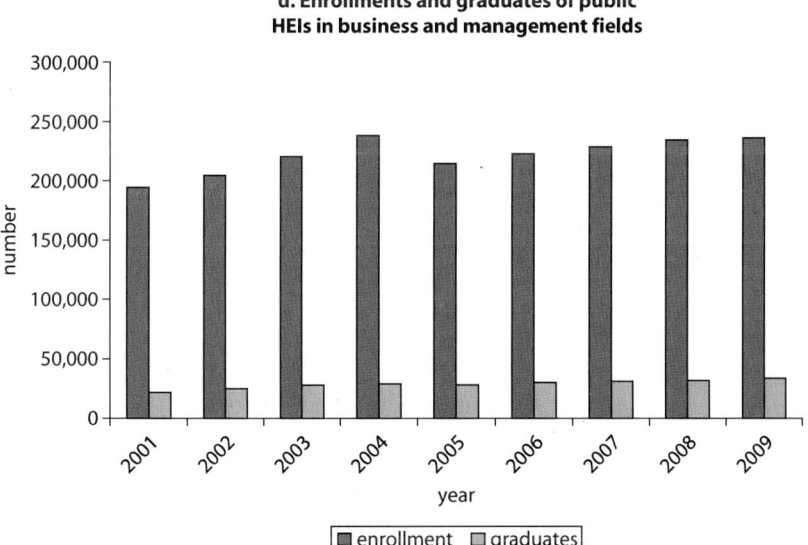

d. Enrollments and graduates of public HEIs in business and management fields

(continued next page)

Figure 3.1   *(continued)*

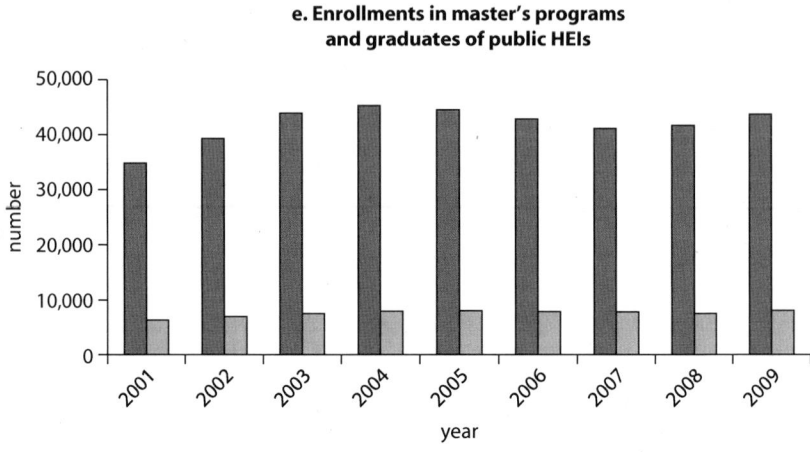

e. Enrollments in master's programs and graduates of public HEIs

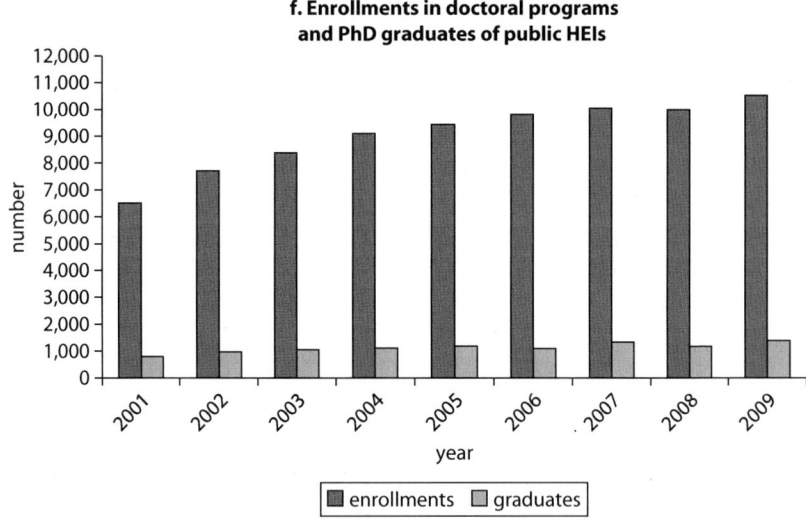

f. Enrollments in doctoral programs and PhD graduates of public HEIs

enrollments   graduates

Source: Government of South Africa, Department of Higher Education and Training, Higher Education Management Information System (HEMIS database).

within five years, with another 15 percent still registered after five years. The remainder, about 70 percent, dropped out without completing. In the case of universities and technikons (a nonuniversity South African HEI, focusing on vocational education), only 30 percent of the students who enrolled in 2000 graduated within five years. Another 14 percent were still registered after five years, and 56 percent left without graduating.

The result of the combination of low enrollment and low completion rates is therefore a low supply of skilled workers. In 2009, the public postsecondary education system added 40,973 science, engineering, and technology graduates; 33,788 business management graduates; 8,112 master's degree holders; 1,380 PhD holders; and an unknown number of FETI graduates to the labor force. In the meantime, the working-age population increased by 0.8 million, from 30.9 million to 31.7 million, whereas the size of the labor force was only 17.5 million.

The Supply-Side Constraints

When a persistent skill shortage generates a significant premium for skills in the labor market, as indicated by the growing disparity between wages of skilled and unskilled labor, a well-functioning postsecondary education system would have responded with a dramatic expansion of investment in capacity and an increase in supply. That this has not happened in South Africa suggests the existence of serious constraints on the supply side. Identifying such constraints, and establishing causality between them and the observed lack of supply-side response to the skill shortage, needs much more intensive investigation into the data than the scope of this book allows. Nonetheless, existing evidence does allow a few observations with varying levels of confidence.

The first constraint, which is well recognized and beyond any doubt, is the quality of school education, which has resulted in uneven preparedness among entrants to the public higher education and further education and training systems. Known as the *articulation gap*, a mismatch results between the entry assumptions of those entering higher education and the actual competencies gained by these graduates. Poor quality of school education is widely regarded as the main cause behind the low completion rates, as shown in figure 3.1 (Government of South Africa DHET 2011, 37). South Africa is particularly lagging behind the rest of the world in mathematics and science education. The 2009 *Africa Competitiveness Report* ranked South Africa 132nd of 134 countries in terms of the "quality of mathematics and science in the country's schools" (WEF 2009, 227). Clearly, the fundamental improvement in skill supply in South Africa cannot happen before a turnaround of the quality of school education, particularly in math and sciences.

Human resources constraints are clearly responsible for the inadequacy of the skill supply in South Africa, as implied by the data presented in figure 3.1. The inability to attract and retain academic staff and the replacement of an aging, predominantly white professoriate and research community are matters of growing concern. They were cited as one of the main challenges facing higher education in South Africa (Stumpf 2010, 39). The stagnation of teaching and research capacity in public HEIs and FETIs is partly a reflection of the skill shortage itself, because qualified professionals have been in short supply as well. One factor behind this problem is the "brain drain": outflow of South African talents to foreign countries for various reasons.

Data availability does not allow an investigation into other factors, but financial constraints, inadequacy of infrastructure, poorly developed links with business and industry, limited scope of private sector participation, and internal governance weaknesses of public institutions are all likely responsible for the dismal supply-side response to skill demand.

Strategic Direction for Reform

South Africa needs to take urgent action on a very large scale to dramatically increase the supply of skilled labor. The urgency is partly caused by the path-dependent nature of skill development, because as NEET youth grow up, their employability diminishes and so does the likelihood of reversing the trend. South Africa already has an unemployed population characterized by the lack of any working experience. In September 2010, 65.8 percent of the unemployed had not worked for the past one year, and 58.6 percent had not worked for the past five years (Statistics South Africa 2010). As time goes by, raising the skills of unemployed workers without any work experience will be increasingly difficult and costly, if not totally impossible, and the social consequence of a large population of middle-aged, nonemployable adults could be severe.

The need for large-scale action is justified by the size of the challenge in South Africa. The total number of NEETs and new entrants to the labor market, which could potentially be in the range of 3 million to 4 million people, renders any marginal improvement in the postsecondary education system, with a capacity of about 1.3 million, insignificant. Quickly containing the trend requires decisive action with dramatic policy changes and massive investments.

Although a deeper diagnostic study based on richer data is still required, existing evidence suggests that a dramatic increase in skill supply would entail joint efforts by both the public and the private sectors. A reform strategy oriented toward a public-private partnership, whereby the government concentrates on quality assurance and financing, mobilizing both public and private service providers to increase service provision, may have a better chance of success. In terms of priority, such a strategy could include the following three components.[5]

First, priority should be given to the urgent challenge of NEETs. In view of the severe social consequences of the large number of NEETs, the advice is for the government to consider extraordinary short-term actions to provide some form of learning opportunities to NEETs, if formal education, employment, and training are not possible. This intervention could serve as a temporary measure for addressing the poor quality of school education[6] as well as alleviating the social pressure of youth unemployment.

Second, capacity should be rapidly scaled up for TVET in general and workplace learning in particular. Though the capacity of the higher education system also requires dramatic expansion, especially in the science, engineering, and technology fields, priority should be given to the TVET system and, in particular, workplace learning. Conceivably, many more postschool youth would find entering the labor market through either the FETI route or an apprenticeship or learnership easier than through higher education, making workplace learning even more critical to their success, which is also a major weakness of the existing TVET system.[7] In particular, emphasis should be given to demand-side training and the need to reinstitute apprenticeships within the training system. The National Tooling Initiative introduced by the Department of Trade and Industry is one such program that has been introduced to revive the apprenticeship model, working in close partnership with FETIs to support and revive the tooling industry. Other examples of successful programs in the past include a public-private partnership initiative set up in the business process outsourcing sector in the Western Cape that has trained close to 7,000 professionals in the past four years and placed nearly 70 percent of those graduated in permanent positions in the industry. Another example is the case of the boat-building industry in Cape Town, which needed specific training for the niche luxury yachts that it built. The government provided funding to the technikons to import machinery that is more specific to the training needs of the industry.

Third, South Africa also needs to strengthen its position in the international competition for talents. Once a sound financing regime is in place, the recruitment of qualified educators will be critical to capacity expansion. Because high-end skills are characterized by greater international mobility, South Africa will need to participate in the international competition for talents more effectively to reduce "brain drain" and maximize "brain gain." Concerted efforts are needed to sufficiently strengthen South Africa's position and attractiveness for the competition.

The Skill Development Challenge in Other Southern African Countries

Similar conditions affect the skill supply in Mauritius. The findings in this section on the Mauritian education system and the skill training levy are derived from the analysis conducted in a parallel study on skills and technology absorption in Mauritius (World Bank 2011b). Similar to the case of South Africa, student attrition of 31 percent is a major concern, and only 13.5 percent of total primary school entrants pass the school certificate, which is very low compared with 90 percent of the students graduating from primary to secondary education in Mexico, Turkey, and Vietnam (World Bank 2011a). This high attrition rate after primary education contributes to a low skill base in Mauritius, with students as young as 12 years of age who are not successful in passing the certificate of primary education after two attempts being forced to attend prevocational training outside the general education stream. The prevocational stream is not geared toward teaching the student core literacy skills, making it extremely hard for students who are in the prevocational stream to pursue higher studies or integrate into the labor market.

Although total tertiary education enrollment has tripled over the past decade, the rise is to a large extent because of the large increase in the number of students going overseas. The large number of secondary graduates leaving to study abroad is indicative of the limited capacity of the current Mauritian higher education system to cater to the demands of these young graduates. Higher education also receives a very small share of the total education expenditure outlays. The allocation of public expenditure outlays is heavily biased toward primary and secondary education, making up almost 75 percent of the total. This level of expenditure is in stark contrast to another small island economy, Singapore, which allocates only 46 percent of its total public education expenditure to primary and secondary education. Mauritius is also facing a skill shortage in the teaching community, with the percentage of PhD holders among the academic staff increasing to 40 percent in 2007–08, though that is still low when compared with international standards. Modest salaries paid to academic staff hinder attracting highly skilled Mauritians or foreign academics to join the universities.

Furthermore, the TVET system has increased rapidly, with a lack of coordination between the private providers leading to fragmentation and a certain amount of duplication. Though a high level of dialogue between the providers and the private sector occurs (unlike the case of South Africa), the current TVET programs are not designed to meet the demands of the labor market. An analysis of the training levy suggested that only a small percentage of firms is using the training levy, with the number of firms contributing to the levy far surpassing the number of firms using the training grants.

Given that Mauritius aims to be moving toward a knowledge-based economy, demanding new higher-end skills, a fundamental restructuring of the current education system is called for to meet the skill demand. The age of students who get into prevocational training could be increased, they should be equipped with the requisite learning skills for the labor market, and allocation inefficiencies should be tackled. The TVET system should be realigned to cater to the changing demands of the economy, with greater emphasis on new sectors not catered to in the current system, such as information and communication technology, and to provide a greater number of technicians and professional skills. Reforms are also needed in the training delivery sector, including provision of adequately trained instructors and better-equipped facilities. In the higher education sector, similar reforms would include curriculum revisions to better align with the needs of emerging industries and to improve the quality of academic staff. Because those reforms would have implications for public debt, the introduction of income-contingent loans, whereby students pay back loans in relation to the income they derive, could be explored.

The Mauritius case is different from South Africa's in that Mauritius's skill shortage is partially alleviated by permitting firms to hire expatriate skilled professionals. An occupation permit allows an eligible noncitizen to live and work in Mauritius for up to three years. It is both a work and a residence permit. Application for this permit is made at the Occupation Permit Unit of the Board of Investment. A complete application is determined by the Passport and Immigration Office within three working days of its date of submission. To qualify for an occupation permit as a professional, a noncitizen should hold a valid contract of employment with a local company or partnership. The basic monthly salary at the time of the launch of the scheme was a minimum of MUR 30,000 (US$1,000), which was increased to MUR 75,000 in 2009 and has been reduced to MUR 45,000 in the 2011 budget speech.

Although Mauritius is able to alleviate some of its skill shortage by hiring expatriate labor, given the perceived skill shortages, Mauritius still needs to introduce measures to increase the scale and quality of its workforce. On the supply side, investing in science and engineering education to strengthen Mauritius's technical workforce should be a government priority. Collaborating with overseas universities (for example, in the United Kingdom and Australia) could be an additional channel for effective human resource development. Foreign universities can help develop and upgrade curricula and teaching materials as well as provide teaching staff.

These collaborative initiatives could be undertaken by tapping into the Mauritian diaspora. Ongoing efforts initiated by the government of Mauritius could be further strengthened to attract skilled Mauritian researchers and academics to contribute to and encourage collaborative programs between Mauritian nationals abroad and research institutes and universities in Mauritius. These efforts could include taking Mauritian students into their labs or research institutes and providing them with lectures when they return home to visit their families. Taiwan, China, offers an example of an effective program. The government's National Youth Council was able to formulate policies in the 1970s that connected Taiwanese businesses with skilled migrants, which allowed them to garner synergies from those migrants abroad. The council tracked migrants in a database, advertised jobs overseas, and provided travel subsidies and temporary job placement to potential returnees. Moreover, the National Science Council and Ministry of Education recruited thousands of migrants as professors and visiting lecturers for the country's growing universities. The government of Mauritius has initiated the development of a diaspora database. This project should be done as a priority to gather information about Mauritians in Australia, Canada, France, and the United Kingdom.

Addressing the labor market constraints should be a priority in Namibia and Lesotho as well. The ease of importing skilled labor into Mauritius stands in stark contrast to the situation in South Africa and in Namibia, where hiring expatriate labor and obtaining the requisite work permits are extremely difficult. A short-term solution for partially alleviating the severe skill shortage in South Africa and Namibia would be to facilitate the hiring of skilled personnel from abroad. The possibility of relaxing the black economic empowerment rules while skills are scarce should also be considered. In Namibia, expediting the process of obtaining a work permit and granting permits for longer than the current three months should be considered. Labor costs are substantially higher in Lesotho than in comparable Asian countries (R 890 a month versus

R 250 to R 450 per month), especially given the low skill level and high turnover rates. The skill base of the Lesotho labor pool should be expanded. Though the recent establishment of a basic training facility for machinists is recognized as a positive development by the firms surveyed, significantly more needs to be done to bolster the skill profile of the workforce—at both machinist and more technical levels. Other alternatives include providing greater links with the University of Lesotho and addressing the prohibitive costs associated with bringing technical training experts to Lesotho.

Supporting Learning through Trade

The case studies in chapter 2 highlighted the importance of trade as a channel for technology transfer and absorption. Exporters in Mauritius, Namibia, and South Africa faced competitive pressure that led them to invest in technology upgrades to compete in the global market. Links with their suppliers and customers provide them access to the technology they need. The purchase of machinery and equipment was a very important channel of technology transfer. In the South Africa NIS, 80 percent of the firms cite acquisition of machinery, equipment, and software as the primary channel by which they acquire new technology. In Mauritius, 54 percent of firms said the same.[8] These results are corroborated in the case studies, in which a large proportion of the technology transfer in Mauritius, Namibia, and South Africa took place by the acquisition of machinery and equipment from abroad and the accompanying consulting services obtained from the equipment supplier.

What can the government do, then, to support technology absorption through trade? In addition to ensuring general openness to trade and a supportive business environment, case studies in chapter 2 and experience in other developing economies point to the need for government support for learning through trade, including learning from exporting, importing, and knowledge transfer supported by what is known as "technology diplomacy."

Supporting Learning from Exporting

Learning from exporting often takes place when exporting firms are under pressure to meet quality standards, including those regarding safety and environment, established by their customers or the regulatory authorities of the destination countries. Such pressures can either strengthen incentives for exporting firms to upgrade their technology or hinder other firms that lack the minimum requisites from exporting to these more sophisticated markets.

Although competition pressure helps provide the needed incentives, capacity is often the binding constraint that prevents firms from learning more from exporting. The inability to adhere to minimum global standards, in particular, can pose a significant obstacle for firms to enter into new export markets. In Mauritius, for example, firms in the seafood hub that processed frozen seafood had high levels of phytosanitary standards to adhere to, which required constant testing and ongoing R&D. Capacity constraints are particularly severe for small and medium enterprises (SMEs).

Government support may therefore focus on three areas. The first is financial assistance to defray a portion of the cost that the firm would have to incur to acquire the requisite capacity. In Mauritius, for example, the government-sponsored Manufacturing Adjustment and Small and Medium Enterprise Development Fund provided funding support to eligible businesses for product quality, standards, and packaging. Reimbursements were made at the end of the project or in phases, depending on the nature of payment terms. Examples of costs that were eligible under this scheme included consultancy and certification (one-off certification), product testing or conformity assessment (to meet country norms and standards) by local and international laboratories, and design of new packaging and labeling (World Bank 2011b). The Namibian government also offers certain incentives to manufacturers and exporters of manufactured goods, as summarized in annex table 3A.1. At present, the government provides few support programs for the agroprocessing industry and no support for incentives to undertake R&D and invest in new technologies. A matching grant program should be considered to support the training of workers and the purchase of consultancy services, including those required for quality certifications to adhere to the global standards. The firm would benefit from not having to bear the entire cost of the investment, and because the firm matches the portion paid for by the government, the program would attract genuinely committed firms.

The second area of support is public provision of basic infrastructure for technology transfer and absorption. Adequate infrastructure facilities to enable firms to adhere to international standards play a vital role in encouraging exports. This support is especially important for countries that are resource rich and need to produce value-added products for exports. Kenya's prepackaged vegetables and cut flowers, which adhere to phytosanitary standards, command a higher price and are sold in European markets because of the additional processing involved. The government's role in helping firms meet international standards has been instrumental in Kenya's horticulture success.

The analyses in chapter 2 showed a lack of adequate supporting infrastructure for at least some sectors in southern African countries. For example, Namibia had no accredited control laboratories and insufficient enforcement of standards by the government to support firms in the agroprocessing industry, where adhering to global standards is the minimum prerequisite to survival. The government of Namibia could invest in building adequate veterinary services, for example; improvement of the field is key to growing exports in some high-value niche markets. Furthermore, if Namibia joined the Marine Stewardship Council, the fish-processing sector would have access to numerous benefits, including potential increased sales to Marine Stewardship Council member countries, a greater ability to attract capital investments and joint ventures, reduced market volatility, lower uncertainty costs, and a more sustainable fishery.

Third, the government and public institutions have the potential to play a role in brokering relationships, including referral to experts and private consultants; fostering the establishment of business networks and interfirm groups; and supporting supply-chain management (interfacing with original equipment manufacturers and SMEs).[9] This category of activities may also include facilitating the spin-off of new small enterprises from universities and other organizations, the "spin-in" of small enterprises to develop their relationships with researchers and other sources of expertise, the establishment of virtual business networks to foster exchange, and the use of innovation vouchers to introduce SMEs to new sources of public and private sector assistance (Shapira and Saritas 2010).

Supporting Learning from Importing

The process of learning from imports of capital equipment is significantly enhanced when it is accompanied by training and by onsite technical assistance for installation of the new machinery and equipment and associated plantwide production, quality control, and inventory systems. As previously noted, acquisition of a new machine from a sophisticated supplier is often accompanied by assistance in the installation of the machine, provided by engineers of the supplier in the plant location, as well as by training in operation and maintenance of the machine. Thus, because the importation of plants and equipment is the major channel for technological enhancement, countries should facilitate and encourage this process, along with the accompanying technical assistance and training activity.

Although few barriers to importation of capital equipment were found in the studied countries, the case studies in this book highlighted multiple examples of obstacles faced by local importing firms that prevented their access to the training expertise provided by plant and equipment suppliers. In Namibia, for example, firms complained that work permits and visas were extremely hard to obtain for engineers and other supplier personnel, thereby making installation inefficient and severely limiting any accompanying learning. In Mauritius, a pronounced tendency existed to buy secondhand machines or low-cost machines from unknown suppliers, which generally will not result in assistance in installation or training. Though this approach may be advantageous and cost-effective in the short run for the importing firm, the long-term benefits of enhanced skill acquisition and training may have been lost.

Support that is targeted at training will enhance technology transfer through the importation of plants and equipment and simultaneously enhance training that accompanies importation. This benefit will, in turn, raise labor productivity and hence possibly generate employment.

Countries in this study would benefit from investigating the possibility of having a scheme whereby a firm that imports new or secondhand machinery can get support for a portion of the cost of the training that accompanies the importation of the machine. This approach is effectively a matching-grant scheme that addresses a defined market failure by supporting training that enhances technological performance. This approach would be particularly beneficial for firms that have purchased secondhand machinery and that do not have access to the training and consulting services that normally accompany the purchase of new machinery and equipment.

Developing "Technology Diplomacy"

One form of public support to technology absorption through trade that has received less attention is what can be called *technology diplomacy*, in which governments make use of their bargaining power in trade to promote technology transfer to their domestic economies (UNECA 2010). For example, in 2003, the Chinese government and Microsoft signed a deal to use Windows as the preferred desktop operating system for government offices. In return, Microsoft was required to reveal its Windows source codes to allay the security fears of the government and to cooperate with the country's largest software development and integration firm to codevelop products based on Microsoft's software platforms. Microsoft also trained 200 software developers and 120 architects within one year. Rather than simply allowing Microsoft to wire up the government operations, the contract was clearly designed to promote technology diffusion (UNECA 2010).

Another example is the Airbus-Aeroflot deal involving the purchase of 22 A350 Airbus planes by the state-owned Russian Federation airline in March 2007. This deal included the participation of Russian firms in the production of the planes. A number of components for the production of Airbus planes were manufactured by Russian plants and the Engineering Center Airbus Russia, one of Airbus's design and engineering centers. This deal, estimated to be worth about US$25 billion, followed three partnership agreements proposed by Airbus in 2006 (engineering and manufacturing of parts, conversion of passenger planes to cargo planes, and participation in design and manufacture of new-generation Airbus planes) with Russian firms and government (UNECA 2010).

These are all examples of "market for technology," that is, trading market access for technology transfer. Their relevance, therefore, varies across countries whose bargaining power differs substantially. The examples are nonetheless illustrative of the potential for governments to make good use of their bargaining power to gain access to technology owned by powerful foreign firms. Though smaller and poorer countries often have weaker bargaining power, clear objectives and effective strategies can help make the best use of trade to promote technology transfer.

Taking Proactive Actions to Increase FDI Spillover

To the extent that FDI is an important channel of technology transfer and absorption, the case of the Lesotho textile industry studied in chapter 2 deserves particular attention. Lesotho was one of the most successful African countries in attracting FDI to develop the textile industry. In 2004, Lesotho registered the largest apparel export to the U.S. market among all African countries that had African Growth and Opportunity Act (AGOA) status (Portugal-Perez 2008). However, the case study indicates little success in technology transfer and absorption through spillover to the local economy. The case, therefore, raises an important policy question: what can be done by countries such as Lesotho to increase technology absorption through spillovers from FDI?

Indeed, Lesotho is not alone. Although developing countries have competed fiercely to attract FDI, numerous empirical studies point to mixed results in terms of positive spillover. Existing empirical evidence shows that the theoretically postulated spillover effects do not materialize automatically just because a country is able to attract FDI in the first place (Reis and Farole 2010). However, no generic formula leads to FDI spillovers in the host country, which makes answering the policy question raised by the Lesotho case in a generalized way difficult. Nonetheless, more successful experiences, especially in the textile industry, do provide important insights for policy makers to consider.

Incentives of FDI Firms to Transfer Technology
Technology transfer and absorption do not happen automatically just because of the presence of FDI firms. Various barriers keep firms in the host economy from the technologies owned by FDI firms, and FDI firms would incur certain costs if they engaged in transferring technology and skills to the host economies. Incentives are therefore important for FDI firms. Absent such incentives, the behavior of the foreign investors with regard to technology transfer will be dictated completely by their commercial calculations.

The case study on the textile sector in Lesotho shows that both the South African and the Taiwanese plants in Lesotho have been set up as low-cost assembly operations in support of well-established, higher–value-adding design and development facilities located elsewhere. This arrangement takes advantage of preferential market access into the United States (for Taiwanese-owned firms) and fewer labor regulations (for South African–owned firms). The case study found no incentive on the firms' side to transfer knowledge and skills to the local economy beyond the point of their own need.

The clothing industry is dominated by powerful organizational buyers with global sourcing strategies. To access this trade, most countries have integrated themselves into FDI value chains headquartered in Hong Kong SAR, China; Taiwan, China; and the Republic of Korea, offering either low labor costs or, in the case of Lesotho, preferential market access. The disadvantage of this approach, however, is that higher-value functions are confined to the headquarters, limiting upgrading. To a degree, Bangladesh and Mauritius have managed to avoid this low-technology trap, in part because they entered the industry much earlier when barriers were lower and the MFA was still in force. Unlike Lesotho, however, they also had other elements working in their favor. Staritz (2010) highlights the importance of entrepreneurial tradition in Mauritius and government support for small business in Bangladesh (with respect to the bonded warehouse and loan and credit facilities). In today's highly integrated global value chain, incentives of FDI firms to transfer technology and skills are more important than ever before.

The experience of Thailand, as studied by Khan (2009), provides a similar story. Since the end of the Asian financial crisis, Thailand has become the regional hub for sophisticated electronics products such as hard drives and has attracted a significant number of Japanese car manufacturers, including Mitsubishi and Toyota, which are increasing capacity and consolidating Thailand as their premier regional base. However, Thai ownership in the advanced segments of these markets has been almost entirely driven out. The very policies that attracted multinationals to Thailand after 1997, including the free trade agreements and the credible protection of intellectual property rights, also kept domestic producers from continuing with their strategies of building technological capabilities. No specific policies were in place to create incentives for foreign investors to transfer technological capabilities. So while foreign investment came in from the 1980s, much of it was to use Thailand's existing knowledge base to achieve production for global markets. The indigenous Thai entrepreneurs and independent higher-level technical personnel were increasingly left out of this process. The integration that took place was between the Thai operations of multinationals and their own global supply chains. Although some learning and capability development took place, no special incentives were provided to the foreign investors to invest specifically in the local economy through skill development or technology transfer. Hence, operations were completely determined by the commercial calculations of these multinationals.

In clear contrast is the case of Bangladesh. In 2005, Bangladesh's garment exports amounted to US$6.9 billion, which was about 2.5 percent of the global total (US$276 billion) and 75 percent of the country's exports (Haider 2007). Exports grew at double-digit rates for more than two decades, as shown in table 3.1.

The success of Bangladesh's garment industry is one of FDI spillover. The take-off of the industry started in 1979 when a Korean company, Daewoo, signed a five-year collaboration agreement with Desh Garments, a factory created by a Bangladeshi entrepreneur. Under this agreement, Daewoo received 3 percent of Desh sales in return for aid in marketing products, purchasing machinery and fabrics, and training Desh employees. The training included technical, management, and

Table 3.1   Bangladesh's Garments:
Growth Rates of Dollar Exports, 1985–2006
percent

Year	Woven	Knitwear	Total garment exports
1985–90	n.a.	n.a.	45.9
1990–95	n.a.	n.a.	24.1
1995–2000	n.a.	n.a.	14.3
2000–01	n.a.	n.a.	11.7
2001–02	−7.1	−2.5	−5.7
2002–03	4.3	13.3	7.2
2003–04	8.6	29.9	15.8
2004–05	1.7	31.3	12.9
2005–06	13.5	35.4	23.1

Source: Khan 2008, table 21.
Note: n.a. = not available.

marketing skills, and Daewoo technicians were sent to Desh to help set up machines and oversee quality assurance (Staritz 2010). On June 30, 1981, Desh prematurely terminated its contract with Daewoo, taking off on its own. Inspired by Desh's success, Desh's employees left Desh to set up or join other garment factories that were springing up. Two rounds of diffusion helped replicate Desh's success in other factories, leading to the phenomenal growth of the industry. By 1985, just five years after it all started, 700 garment export–manufacturing factories were operating in Bangladesh (Khan 2009). What brought Daewoo to Bangladesh was the rent created by the Multifiber Agreement on Textiles and Clothing (MFA), which was similar to the rent created by AGOA in the case of Lesotho, which attracted the Taiwanese investors. However, in the case of Bangladesh, extra incentives, namely, 3 percent of Desh's sales, were provided to Daewoo, and perhaps most important, they were specifically linked with its performance in knowledge and skill transfer.

Similar success occurred in Mauritius. The Mauritian clothing and textile industry also "started" in the 1970s, a few years earlier than Bangladesh, when foreign (mostly Hong Kong SAR, China) investors set up plants that assembled imported textile inputs into clothing. Much of this activity took place in export processing zones (EPZs), where investors received tax benefits, duty-free imports of raw materials, and machinery in return for agreeing to export all their products. However, Mauritius managed to establish backward linkages into textile production, and with the phasing out of the MFA, local producers took the place of Hong Kong SAR, China, firms in the production chain. In the 1990s, Mauritian firms, particularly the ones exporting to the European Union, focused on improving product development and design capabilities and on diversifying buyers.

Another successful case in which foreign investment played a vital role was Kenya's horticulture industry. A vibrant, innovative private sector found ways to produce new horticultural varieties that were preferred by foreign consumers and, most important, gained the reputation of meeting quality, environmental, and labor standards, and thus gained niche markets and greater profit margins. The main sources of expertise were foreign investors and partners who advanced production technologies. The role of the government is cited as encouraging the initial FDI and working with the private sector to help the enterprises meet international standards (Watkins and Ehst 2008). The success in this sector was private sector driven. Singapore and Ireland also offered foreign investors a variety of incentives while simultaneously investing in training and infrastructure, thus buttressing public investments with incentive mechanisms to attract domestic and foreign investments (Yusuf and Nabeshima 2011).

Incentives of Local Firms to Learn

The Bangladesh case is also one of successful learning. As Khan (2009) observed, Bangladesh would not have succeeded if it had simply produced garments using existing capabilities under the protective barrier created by the MFA. The success was made possible by the acquisition of knowledge by local firms. Evidence of this conclusion lies in the fact that once the new production routines had been established and understood, many new garment factories opened in Bangladesh within two or three years. So dramatic was the growth that the United States imposed quotas on Bangladesh in 1985, just a few years after Desh Garments actually began exporting. By the time the MFA was effectively removed in 2007, Bangladesh had a diversified garment industry that was engaged in some backward and forward linkages as well.

According to Khan (2009), a number of conditions help explain the successful acquisition of technological capabilities by Desh and its transfer to the emerging Bangladesh garments sector. First, all the local companies had to make relatively uncertain investments in learning to produce different types of garments to global standards, despite the rent created by the MFA. "Cofinancing" in this form meant that the investor had very strong incentives to put in high levels of effort, because the quota rents were not sufficient to achieve global competitiveness with the existing levels of capabilities. Second, the allocation of the MFA rent was such that everyone knew not to expect the MFA to be a permanent feature of the global economy. Third, the joint venture between Desh and Daewoo involved significant investments from both sides, but particularly the Bangladesh side, which had to put in most of the effort. The founder of Desh, Nurul Quader Khan, provided the bulk of the capital required for the investment in plant and machinery for the venture. Daewoo's investment was to invite a large team of Bangladeshis to visit and study its operations in Korea and to understand by being onsite how production was actually set up and organized. The characteristics of this firm-to-firm relationship and the internal incentives and compulsions within Desh are perhaps the most critical characteristics explaining the success of the technology transfer, according to Khan. The fourth factor is that the sector had relatively small-scale economies and motivated managers could therefore expect that if their learning was successful, they could one day be entrepreneurs themselves. Of course, the nature of the technology was important for sustaining those incentives, as well as the size of the market. By the end of the 1980s, of the 130 people who were first trained by Desh in Daewoo's factories in Korea, 115 became entrepreneurs and set up their own garment firms. This competition apparently did not do much damage to Desh, whose output and profits continued to grow unabated.

Government support was also crucial. As other Korean firms started investing in Bangladesh, local residents came to perceive the clothing sector as lucrative. Government policies such as bonded warehouses, back-to-back letters of credit, and cash incentives helped locals capitalize on this investment and start their own firms. With back-to-back letters of credit, exporters were able to open accounts in a local bank for the import of inputs against the export orders placed in their favor by the final clothing importers. The cost of the imported items, along with interest and other charges, would be deducted by the local bank from the proceeds of the sales of the final output. Benefits such as duty-free imports were available to all clothing firms. Bangladesh was further aided by very low labor costs and economies of scale (Staritz 2010). Bangladesh's main competitive advantage is its low labor costs, likely the lowest in the world. It has a comparatively long experience in the sector and capabilities in addition to cut-make-trim.

Interestingly, the geographical location of domestic firms in the industry is such that only a few benefit from preferential policies in the EPZs.

The current Bangladesh industry is highly geographically concentrated, with 75 percent of firms located in Dhaka and the rest mainly near Chittagong. Like Mauritius, Bangladesh has made use of EPZs. Although about 65 percent of the foreign-owned firms are based in these EPZs, the zones account for less than 1 percent of the total industry. The majority are based outside the EPZs, forfeiting better infrastructure and tax rates in return for lower production costs.

Proactive Actions in Technology Absorption

Technology transfer and absorption do not happen automatically even when a country is successful in attracting FDI. The process, therefore, takes more proactive actions than generally improving the business environment to turn the success in attracting FDI into one of technology transfer, absorption, and growth of the indigenous private sector. Still, the first-order challenge in southern African countries remains to attract more FDI. South Africa, in particular, seems to be lagging behind its Asian peers in attracting FDI into manufacturing industries (World Bank 2010, 24). However, this indeed makes imperative the need for southern African countries to maximize the benefit of spillover from the existing FDI stock. Proactive actions in technology absorption are therefore complements to, not substitutions for, actions to attract more FDI. Three sets of proactive actions to maximize FDI spillover can be considered.

The first is to provide incentives to foreign investors. Almost all countries have something to offer to foreign investors, such as locations, markets, labor costs, and tax incentives. All these can be used as incentives to gain access to technology and other knowledge owned by the foreign investors and to motivate them to engage in deliberate actions of technology transfer to the local economy. Clearly, in a world where foreign investors, especially multinationals, have many locational options, the bargaining power of host countries is limited. But the Lesotho and Thai cases are instructive in that a strong case can be made for policies that combine incentives for foreign investors with conditions that might significantly improve capability development at higher levels (Khan 2009). Without progress in technology and skill transfer, host economies are likely to be subjected to the risk brought about by locational decisions of foreign investors. Although the first-order challenge remains to attract more FDI, such incentives need to be designed to fully compensate the extra cost that FDI firms would have to incur in engaging in technology and skill transfer. In other words, they should be business incentives, not regulatory requirements.

One critical lesson learned from the Bangladesh experience is that incentives offered to foreign investors should be tied to performance in technology and skill transfer in addition to the amount of the investment or the number of jobs created. Another lesson is that such incentives do not have to be provided by the government. They can also be provided by indigenous private firms with appropriate government support, which can be more effective.

The second set of proactive actions is to put in place incentives for domestic firms' learning efforts. These actions obviously involve the development of a domestic private sector as the "receiver" of foreign-owned technology and skills in the first place. For example, Staritz (2010) highlights the importance of entrepreneurial tradition in Mauritius to its success in developing a local textile sector based on FDI. However, this second set of actions also involves policy actions that would affect the payoff function of potential learners. The government support provided by the government of Bangladesh, as highlighted previously, is one example that aids domestic firms to capture the gains from learning by competing or collaborating with foreign firms. In countries where such receivers are in short supply, government investment in supporting domestic firms to effectively collaborate with FDI firms may have high long-term returns. Effort needs to be made to build on the existing FDI platform to grow the local industry, by fostering greater links between foreign and domestic investment, such as supplier development programs that encourage domestic firms to supply to foreign firms. Moreover, greater information dissemination is needed on the existing incentives provided by the Lesotho National Development Corporation, which are available to foreign as well as domestic investors, because the general perception is that these incentives are only available to foreign investors.

The third set of actions involves measures to be taken to strengthen absorptive capacity. Skill development, which has been discussed earlier in the chapter, is an obvious priority. The development of basic infrastructure for technological progress, such as the system of metrology, standardization, testing, and quality, as well as technical advisory services, is another potential priority. R&D activities are generally instrumental in enhancing capacity for technology absorption. The strategy could identify a set of R&D activities that are potentially crucial to the absorptive capacity of domestic firms and provide targeted support. Lesotho, for example, needs to take advantage of the regional programs that exist and tap into the existing research base in South Africa. Finally, these efforts need to be coordinated within the National Strategic Development Plan that the government of Lesotho is currently formulating.

The Need for Diversification in Lesotho

The lack of success in technology absorption has imposed challenges to economic policies in Lesotho because of its dependence on FDI-sustained textile exports to the U.S. market. Given the eventual phasing out of preferential trade access and the increasing number of low-cost competitors in textiles, FDI aimed at the South African market may be a more beneficial focus for public technology and innovation policy in Lesotho.

The comparison with Mauritius and Bangladesh also provides some ideas. Central bonded warehouses could be established in Lesotho to stock up on inputs such as fabrics and dyes, machinery, and spare parts (World Bank 2005). Productivity could be enhanced by moving toward higher-value production, potentially with the support of government funds that complement firm expenditure on new equipment. Lesotho could move some production into "green" manufacturing, such as recycling products derived from the existing industries, for example, cloth cutoffs. These could then be promoted as part of environmental and social standards upgrading. Lead times also need to be reduced, potentially through regional sourcing and a diversification of export markets (Staritz 2010).

This last point is particularly important. Buyers in the developed clothing market are becoming more demanding, preferring to focus on a few strategic suppliers and requiring high levels of manufacturing and services (in terms of product development and design, inventory management and stock holding, logistics and financing, and communication and merchandising). Given the intense competition, only the most capable country suppliers will thrive.

Logistics cost is another area worth addressing, potentially through streamlining the customs process and working with South Africa to coordinate the movement of goods to and from the port of Durban. On a microeconomic scale, potential policy levers could include the institution of benchmarking programs and robust quality management systems. Some form of incentives to promote and develop Basotho entrepreneurs might also be considered. Although the problems facing the Lesotho market are severe, these measures may help create a competitive niche for the Lesotho textile industry. In particular, Lesotho could attract more South African FDI to offset any Taiwanese closures and focus on medium-skill production servicing the regional market. This more dynamic and skilled production has far greater potential for innovation and technology diffusion. Given Lesotho's vision to diversify away from textiles, investment in basic infrastructure is pivotal, or else Lesotho will lose out to other competing destinations. The strategy that the Lesotho National Development Corporation is drafting needs to emphasize the importance of attracting FDI that can transfer knowledge, focusing on long-term potential for Lesotho and not just short-term gains.

Building Absorptive Capacity through R&D and Industry-Research Collaboration

R&D activities are not only crucial to "new-to-the-world" innovation; they also have a critical role to play in "new-to-the-firm" technology absorption because R&D activities, especially those at the firm level, result in stronger absorptive capacity. Cohen and Levinthal (1989) refer to this as the "second face" of R&D. The second face of R&D enhances the ability of firms below the technology frontier to absorb technology to allow speedier catch-up. Therefore, support of R&D and industry-research collaboration is of strategic importance for southern African countries to build absorptive capacity of firms, despite the scarcity of R&D resources.

Strengthening Financial Incentives for R&D

Data from the South Africa NIS provide evidence of the association of technology absorption and R&D. R&D is positively and significantly correlated with the introduction of new processes, new products, and changes in organization. Also, the Enterprise Survey analysis for Mauritius showed that firms that undertake R&D investment had a significantly higher probability of introducing new and improved products (Kuriakose, Goldberg, and Kaplan 2009). However, a high proportion of the firms surveyed identify the lack of funding and the costs associated with R&D as the key factors that significantly constrain R&D spending.

The South African government offers tax incentives in terms of section 11(d) of the Income Tax Act No. 58 of 1962, but some important limitations apply. The program was introduced in November 2006. The incentive consists of a deduction of 150 percent with respect to current expenditures on eligible scientific or technological R&D and an accelerated depreciation of capital assets (including buildings) used for purposes of scientific and technological R&D over three years at the rate of

50:30:20. This rate commences from the year of assessment in which the asset is brought into use. Qualifying scientific and technological R&D activities are "the generation of new knowledge; the discovery of novel, practical and non-obvious information of a scientific or technological nature; or the devising, developing or creating of any invention, design or computer program of a scientific or technological nature" (Government of South Africa DST 2007, 1). The grounds for qualification are broad, and the extent of the credit is significant. However, although precise data are lacking, the take-up, thus far, clearly has been very limited. Very few of the firms interviewed had made use of the incentive. Those firms that have accessed the R&D credit are overwhelmingly large and well established. Though further study is needed, at least three reasons for this trend can be noted. First, R&D tax claimants are required to complete very detailed forms and to provide evidence to substantiate the activities claimed. Second, applicants are required to "indicate the records of publication, patents or similar modes of protection of intellectual property, including plant breeder's rights or designs" (Government of South Africa DST 2007, 7). Third, applicants have been deterred by the difficulties entailed in satisfying the Department of Science and Technology as to the validity of their claims. Only the large firms have the resources and the incentive to persevere. Many expenditures of an applied nature do not qualify, such as prototyping and design engineering. This reason is cited by many firms in the capital equipment sector to explain why they did not make application for the R&D incentive. The inherent limitation of the R&D tax credit is that it can be an effective incentive only where profits are being generated. The credit is of very limited consequence to any R&D activity that is prerevenue and preprofit, as would be the case for newly established companies.

The following modifications can be made that provide incentives to small and medium firms to undertake greater R&D investments: (a) restructure the R&D tax incentive to make access easier, particularly for small firms; (b) extend the list of qualifying expenditures to include more applied R&D, as appropriate, for supporting technology absorption in sectors such as capital equipment; and (c) allow a carryforward of the tax deduction to provide an incentive for R&D activities whose returns do not materialize within one year.

In an attempt to step up indigenous innovation, China undertook a similar reform of its tax administration regarding R&D expenditure in 2008 (China Administration of Taxation 2008). The reformed system requires enterprises to submit only the following six sets of documents: R&D project proposal and budget, names of the members of the R&D unit or task force, consolidated tables of the actual expenditure of the R&D project, documents recording decisions of the board and the CEO on the launching of the R&D project, relevant contracts and agreements, and documents showing the results and impacts of the project. Eight categories of expenditures are eligible: cost of books and translation, staff salaries and materials, depreciation, patent fees, cost of pilot production, tests, inspection, and review. Carryforward of the tax-deduction credit was allowed, and 150 percent of R&D expenditure was made tax deductible, up from the previous ratio of 100 percent.

The traditional approach to R&D and innovation support to firms has been through concessionary loans (and indirect support through tax breaks). However, since the 1980s, OECD countries have been increasingly aware of the benefits of matching grants. The grants have been successfully used in developed countries—such as the United States' Small Business Innovation Research Program, Finland's Funding Agency for Technology and Innovation (Tekes), Israel's Office of the Chief Scientist, and Australia's Commonwealth Scientific and Industrial Research Organization. The companion World Bank study, "Closing the Skills and Technology Gaps in South Africa" (World Bank 2011a), proposes to increase financial support to early-stage innovation through matching grants for product development in its broadest sense, including prototyping and business development.

Other middle-income countries have deployed similar policies. India's Society for the Promotion of Rural Environment and Development is an early-stage technology development program that has been directed exclusively at private enterprises, with an explicit requirement for collaboration with public research institutes, and has been independently evaluated as successful. In Russia, the Fund for Assistance to Small Innovative Enterprises successfully supports small, innovative companies through open competition, providing 50 percent of the project cost.

Similar programs are already in existence in South Africa. The Support Programme for Industrial Innovation in South Africa targets product and process development that represents a significant technological advancement and has a commercial advantage over existing products. Assistance of between 50 percent and 85 percent of the actual direct development costs is granted, up to a maximum grant of R 1 million (about US$150,000). Other mechanisms can also be considered (table 3.2).

Table 3.2   Direct Instruments for Supporting Business R&D

Instrument	Advantages	Disadvantages
Tax incentives for R&D	Provides functional intervention, not picking of winners Offers less distortion, more automatic Generally requires less bureaucracy to implement, although advisable to have monitoring and spot checks	Has unclear fiscal costs in advance, which could be high Is difficult to ensure that R&D increase is induced by tax incentives (additionality) Is not very relevant for start-up firms that do not yet have taxable revenue streams Is a blunt instrument; cannot target specific companies, although it can target specific sectors
Grants for R&D projects	Allows specific targeting on case-by-case basis Can control amount of subsidy granted Can be given in tranches against defined goals Can be structured as matching grants that may help improve quality or efficiency	Requires large bureaucracy to administer May not select the best project Is also difficult to ensure additionality
Accelerated depreciation for R&D equipment	Reduces the capital costs of R&D projects	Does not provide incentives for noncapital costs such as personnel and material inputs
Duty exemptions on imported input into R&D	Reduces cost of world-class inputs if country otherwise has high import duties	Results in loss of tariff revenue; is distortionary to the extent that it favors R&D over other activities
Venture capital to facilitate commercializable research results	Helps overcome financial market failure in making capital available to start-ups with no collateral or track record	Requires detailed knowledge of sectors to evaluate technical and commercial prospects Is often not successful because of limited deal flow and shortage of technoentrepreneurs Also requires developed stock markets so investors can sell off shares and reinvest in new projects

Source: World Bank 2010, table 5.9.

Increasing Industry-Research Collaboration

Collaboration between firms and R&D institutions, including tertiary education institutions and publicly funded scientific institutes, which are the primary location for the production of scientific knowledge, appears to be very limited in southern Africa. In the South Africa NIS, only 11 firms of the 600 surveyed (1.8 percent) rated their links with such institutions as important to their innovation activities. In the interviews, a similar picture emerged—very few firms had any engagement with these institutions, and where such an engagement existed, it tended to be very limited. A decline is seen in the few cases where significant engagement occurred in the past, as in the South African mining equipment industry.

Although limited industry-research links are not uncommon in other regions or countries, 1.8 percent is a rather low level of collaboration. The Community Innovation Survey data show that 8 percent of the firms in the EU-27 countries have links with universities and technikons, and 5 percent have links with government research institutes. These numbers for the United Kingdom are 10 percent and 8 percent; for Finland, 33 percent and 26 percent; and for Poland, 6 percent and 9 percent. In the auto components sector, collaboration between firms and R&D institutions appears to be much lower than in Turkey.

Mauritius, too, has virtually no collaboration between industry and university or research institutes. Firms considered the existing local research irrelevant to industry needs. In Namibia and Lesotho, again no industry-research collaboration is seen, except that large firms in Namibia accessed universities and research institutes abroad, leveraging their foreign partners' relationships.

The locus of technological innovation not only resides within the boundaries of the firm, but also is the result of an interdependent exchange process between different organizations, such as private firms, universities, research laboratories, suppliers, and customers. The rationale for public stimulation of industry-research collaboration rests on the lack of sufficient incentives for individual firms to undertake uncertain and imperfectly appropriable research at the socially optimal level. R&D cooperation between firms and R&D institutions can correct market failures and can increase the rate of technological progress and diffusion of technological knowledge in industry and among research institutes and universities (Racine and others 2009).

The cases of Finland, Ireland, and Singapore bear important lessons. Yusuf and Nabeshima (2011) report that the three countries, at the very outset, envisaged in their strategy a vital role for university-industry links that led to a circulation of knowledge and of researchers. The universities were viewed as a source of entrepreneurship to help transfer innovation to the business sector.[10] In Finland, the Nokia Corporation took the lead in conjunction with the National Board of Education, the Ministry of Education, and the Future Committee of the Parliament in persuading the Academy of Finland to accelerate the initiative to become a knowledge society by mobilizing universities and public research entities (OECD 2007).[11] The role that Tekes played in Finland's transformation into a knowledge economy has been widely acknowledged. Established by the government in 1983, Tekes has a broad mandate that includes identifying areas for technological advance, coordinating the working of the innovation system with catalytic funding of R&D, all the time working closely with government agencies, universities, firms, and private financiers.

In South Africa, the Technology and Human Resources for Industry Programme (THRIP), managed by the Department of Trade and Industry, serves both to promote training that is appropriate for firms and to enhance industry-university collaboration. In its review of the South African science and technology system, the OECD commented very favorably on THRIP: "THRIP has been very effective in integrating the development of research-capable human resources with industry-university co-operation in R&D, and the program has been recognized internationally as particularly successful when compared with similar schemes in other countries" (OECD 2007, 6).

THRIP supports projects on a cost-sharing basis. In 2008–09, THRIP contributed R 139 million (about US$20 million), and industry contributed R 227.5 million (about US$30 million). THRIP has increasingly sought to support small, medium, and micro enterprises (SMMEs) and black economic empowerment firms. Some 40 percent of the university-industry links fostered through THRIP are still operating a year after the end of THRIP funding. In 2008–09, 207 SMMEs and 106 larger firms were engaged in THRIP projects.

This number is significant but still rather small. Moreover, it was reported that the number of SMMEs had fallen sharply, by 22 percent from the previous year. Despite the success of THRIP, the evidence strongly suggests that industry-research links in South Africa remain very limited. A study could be undertaken to assess how THRIP might be extended more widely. Such a study might focus on why firms located in sectors where collaboration would be expected do not engage with THRIP. Other mechanisms to promote links should be considered. For example, one option could be to give priority to consortiums of companies and research institutions, which must be led by the company and not the R&D institute to ensure that the research is market oriented rather than publication driven. The funding flow from the governmental funding agency should go through the company to the research institute and not vice versa.

In Mauritius, the current Mauritius Research Council Private Sector Collaborative Research Grant scheme was implemented with a view to encouraging industry-research links. The scheme gives a grant to an industrial research partnership to undertake commercially viable research. However, analysis has shown that private sector participation in these grants has been limited. The Mauritius Research Council grant should be redesigned to have the private sector in the driver's seat. Research is to be conducted jointly by a private sector company and a research institution, which is restrictive in that it excludes the intellectual know-how and technology transfer that a foreign firm or a foreign university can bring into the mix. Multinational firms or foreign universities (both of which are important sources of technology transfer) should be allowed to participate in these collaborations, thus widening the knowledge pool from which both firms and universities could draw (World Bank 2011b). Furthermore, all three countries—Lesotho, Mauritius, and Namibia—should enable local enterprises to tap into the research capacity in South Africa, in addition to building their own local skills and research capabilities.

Supporting Intracountry Technology Diffusion

Although this book has focused exclusively on technology absorption from the global technology pool under the assumption that advanced technologies are mostly developed in industrialized countries, this focus is by no means the entire story of technology absorption. As indicated in chapter 1, technology transfer and absorption can also take place within one country, one industry, or even one firm. In general, wherever some firms lead others in technological advancement, benefits can be reaped through technology diffusion from those leading firms both within and across industries.

In South Africa, this observation applies particularly to mining and mining-related equipment and specialist services, where a major cluster of South African firms has significant expertise along the global technology frontier. This expertise is evidenced by patent data (World Bank 2011a, chapter 5) and, as has been outlined in this study, by unique South African products and firms that are global leaders in the field. The relative success of mining-related capital equipment exports, linked to South African engineering and project management capabilities, suggests that the spillover potential from this sector in particular needs to be examined and factored into the government's technology objectives.

The challenge is to support the spread of these technologies and companies into new non-mining-related products and markets. To some extent, this expansion is already happening. However, knowledge and information gaps exist, whereby firms fail to see the potential application outside known areas and customers, and new product and new market development does have costs and risks. The fact that firms that move into new areas take risks while much of the benefit of their success falls to follower firms (second movers) constitutes a market failure that potentially provides a space for public policy. The South African capital equipment sector is highly organized and cohesive, with an active export association. Hence, the proposal is that a government task team should investigate, in close consultation with the industry and export association, how firms might be encouraged to apply their technological capacities to new products and new markets.

The spread of frontier-level technologies outside the mining sector into new products and new markets, with the lateral movement of existing technological competencies, should be encouraged. This objective could be considered as a priority area for support in various government programs, such as the 2010/11 Industrial Policy Action Plan and THRIP.

Finland is an outstanding example of a successful country whose growth was heavily dependent on broadening its capacities from natural resource–based industries, principally forest related, into machinery and engineering industries and later into information and communication technology and electronics. The key factors that underpinned this diversification were "the persistent emphasis given to higher education, linkages and spillovers among various industries, and the emergence of new knowledge-based industries" (Dahlman, Routti, and Yla-Antilla 2007, 3). The government adopted a systemic approach to industrial and technology policies. Strong links between research organizations, universities, firms, and industries in relation to knowledge production were promoted, and policy was formulated through public-private partnerships involving economic research organizations, industry federations, and firms (Dahlman, Routti, and Yla-Antilla 2007, 8).

Of course, the Finnish example cannot be exactly replicated in southern Africa. But the approach that is being proposed here—particularly the emphasis on diversification through promoting links and spillovers, a systemic approach to an integrated industrial and technology policy, and the development of policy in close collaboration with the firms and the industry—draws much from the Finnish experience.

Conclusion: Designing Locally Tailored Action Plans

The subject of technology absorption in southern Africa requires further research. The scope of this study has been limited and by no means exhaustive. Case studies have serious limitations when insights gained are generalized. This caveat should be highlighted, together with the conclusions of this study. A carefully designed survey would be a desirable direction for future research. Given the importance attached by firms to the acquisition of machinery and equipment, more efforts should be devoted to looking into details of how technologies are adopted, assimilated, and absorbed through importation of machinery and equipment. The role of the private sector in skill development deserves special investigation where data are available. R&D activities that are oriented to technology absorption should differ from those aimed at technology creation. How they should differ in the context of southern African manufacturing industries is another potential area of focus. Finally, the overall policy direction, and hence options for action, is a subject that requires a great deal of research. Action plans would need to be crafted carefully, taking into consideration the local circumstances, wherever a recommendation from this study is found to be worth trying. Countries, industries, and firms vary dramatically in many dimensions, including the nature of the technology in question. Action plans, therefore, should be tailored to ensure they work in the particular local conditions they are intended for. In this regard, recommendations in this report are more about directions and principles for further policy discussion than about prescriptions.

In particular, in designing new financial policy instruments to enhance technology adaptation and innovation, care should be taken in both the design and management of these instruments to prevent capture or corruption and to promote efficiency. The following elements in designing policy instruments are important:

• Goals and objectives should be clearly established at the outset.
• Performance should be constantly monitored and evaluated against goals and objectives.

- Monitoring and evaluation should be performed externally to the administration and management of the instruments.
- Adjudication to approve applications should be conducted by an independent panel consisting of external peer reviewers.
- The independent panel should have significant private sector and export expertise.
- Adjudication should be based on preestablished, detailed, and transparent criteria.

Annex

Table 3A.1   Incentives for Manufacturers and Exporters in Namibia

Incentive	Manufacturers	Exporters
Corporate tax	Set at 18 percent for 10 years, then reverts to 35 percent	80 percent income tax allowance on income derived from exporting manufactured goods produced in Namibia or not
Value added tax (VAT)	Purchase of machinery and equipment exempted	Normal treatment of 35 percent
Stamp and transfer duty	Normal treatment	Normal treatment
Establishment tax package	Negotiable rates and terms by a special tax package	Not eligible
Special building allowance	Factory buildings written off at 20 percent in the first year and the balance at 8 percent over 10 years	Not eligible
Transportation allowance	Allowance for land-based transportation by rail/road of 25 percent deducted from total cost	Not eligible
Export promotion allowance	Deduction of 25 percent from taxable income	Normal treatment
Incentive for training	Deduction from taxable income of between 25 percent and 75 percent	Not eligible
Industrial studies	Available at 50 percent of actual cost	Not eligible
Cash grants	Covers 50 percent of cost of approved export promotion activities	Not eligible

Source: Kadhikwa and Ndalikokule 2007.

Notes

1. For example, Grant Thornton's International Business Report 2010 reports that limited availability of skilled workers is the major constraint on business expansion in South Africa and is cited as such by 34 percent of businesses in the country. Similarly, 73 percent of South African chief executive officers (CEOs) surveyed in PricewaterhouseCoopers's 2010 *Global CEO Survey* expressed concern about the availability of key skills locally (see http://www.pwc.com/gx/en/ceo-survey and World Bank 2011a).

2. Unless noted otherwise, all data used in this section are from the parallel study World Bank (2011a).

3. http://en.wikipedia.org/wiki/Education_in_South_Africa#Education_under_ Apartheid.

4. Bhorat and Mayet (2011) refer to the presence of skill-biased technical change as the cause of the observed labor demand shift. However, given the marginal role of the manufacturing industry in employment growth (with a contribution of 6.5 percent during 1995–2009), rising skill intensity of employment might be more a result of the lack of development of low-skill-intensive industries than of technology advances. The reason for this gap in skills deserves further investigation but goes beyond the scope of this report.

5. As important as it is, increasing the skill supply from the postsecondary education system is not a substitute for solutions to two other fundamental challenges: the quality of school education and the structure of the economy. The skill shortage problem cannot be completely resolved before major progress is made in improving school educational quality. And the skill shortage will be greater and persist longer for a given supply if the skill intensity of labor demand generated by the economic structure is greater and rises faster.

6. The matriculation completion rate among black and colored youth is less than 40 percent. Because matriculation is the lowest recognized qualification in South Africa, a large segment of the young population is left with nothing to signal its ability on the labor market (World Bank 2011a).

7. For example, in a survey of 9,885 FETI learners in 2009, 44 percent reported having had some form of workplace exposure during their studies, with half of these having been enrolled in a learnership or apprenticeship. The remaining 56 percent had no workplace experience (World Bank 2011a).

8. No data are available for Lesotho and Namibia.

9. The National Research Council of Canada Industrial Research Assistance Program sponsors a network of 240 industrially experienced technology advisers located throughout Canada to assist SMEs in all aspects of innovation, including referrals to qualified consultants. http://www.nrc-cnrc.gc.ca/eng/services/irap/advisory.html.

10. The Massachusetts Institute of Technology and Stanford were the prototypical entrepreneurial universities, and they remain the leaders all others seek to emulate (Yusuf and Nabeshima 2011).

11. This effort was broadened in 2000 with the academy acting in tandem with Tekes, other agencies, and the University of Helsinki to push interdisciplinary research, partnerships, and lifelong learning.

References

Bhorat, Haroon, and N. Mayet. 2011. "Labour Demand Trends and the Determinants of Unemployment in South Africa." Background paper for the forthcoming report on Closing the Skills and Technology Gaps in South Africa, World Bank, Washington, DC.

China Administration of Taxation. 2008. "Tentative Regulation on R&D Tax Deduction." December 10. http://www.chinatax.gov.cn/n8136506/n8136593/n8137537/n8138502/8744184.html.

Cohen, Wesley, and Daniel Levinthal. 1989. "Innovation and Learning: The Two Faces of R&D." *Economic Journal* 99 (397): 569–96.

Dahlman, Carl, Jorma Routti, and Pekka Yla-Anttila, eds. 2007. "Overview." In *Finland as a Knowledge Economy: Elements of Success and Lessons Learned*. WBI Development Studies. Washington, DC: World Bank.

Government of South Africa. 2010. "Cabinet Statement on the New Growth Path." Government Communications (GCIS), October 26, 2010, South Africa. http://www.info.gov.za/speech/DynamicAction?pageid=461&sid=14034&tid=23221.

Government of South Africa, DHET (Department of Higher Education and Training). 2010. Database. http://www.dhet.gov.za.

———. 2011. *Revised Strategic Plan 2010/11–2014/15*.

Government of South Africa, DST (Department of Science and Technology). 2007. "Innovation towards a Knowledge-Based Economy: Ten-Year Plan for South Africa (2008–2018)" (Draft). Pretoria. http://www.esastap.org.za/esastap/pdfs/ten_year_plan.pdf.

Grant Thornton. 2010. *Emerging Markets: Leading the Way to Recovery.* International Business Report 2010. Grant Thornton International.

Haider, Mohammed. 2007. "Competitiveness of the Bangladesh Ready-Made Garment Industry in Major International Markets." *Asia-Pacific Trade and Investment Review* 3 (1): 3–27.

Kadhikwa, Gerson, and Vitalis Ndalikokule. 2007. "Assessing the Potential of the Manufacturing Sector in Namibia." Bank of Namibia Occasional Paper 1/2007, Windhoek, Namibia.

Khan, Mushtaq. 2008. "Vulnerabilities in Market-Led Growth Strategies and Challenges for Governance." Governance School of Oriental and African Studies, University of London, London.

———. 2009. "Learning, Technology Acquisition and Governance Challenges in Developing Countries." U.K. Department for International Development–funded research document.

Kuriakose, Smita, Itzhak Goldberg, and David Kaplan. 2009. "Innovation and Technology Absorption in Mauritius." In *Mauritius: Investment Climate Assessment*, 51–77. Report No. 52794. Washington, DC: World Bank.

OECD (Organisation for Economic Co-operation and Development). 2007. *Review of South Africa's Innovation Policy*. Paris: OECD.

Portugal-Perez, Alberto. 2008. "The Costs of Rules of Origin in Apparel: African Preferential Exports to the United States and the European Union." Policy Issues in International Trade and Commodities Study Series No. 39. UNCTAD, Geneva.

Racine, Jean Louis, Itzhak Goldberg, John G. Goddard, Smita Kuriakose, and Natasha Kapil. 2009. "Restructuring of Research and Development Institutes in Europe and Central Asia" (Draft April 13). Europe and Central Asia, Private and Financial Sector Development Department, World Bank, Washington, DC. http://siteresources.worldbank.org/INTECAREGTOPKNO ECO/Resources/ECAKE3_April_13_09_Exec_Sum.pdf.

Reis, José Guilherme, and Thomas Farole. 2010. "FDI and Global Value Chains in Sub-Saharan Africa: Understanding the Factors That Contribute to Integration and Spillovers" (Draft). World Bank, Washington, DC.

Shapira, Philip, and Ozcan Saritas. 2010. "Supporting the Upgrading of Small and Mid-Size Enterprises: International Experiences and Best Practices from Selected Countries." Draft report prepared for the World Bank, Washington, DC.

Sheppard, Charles, and Nico Cloete. 2009. "Scoping the Need for Post-School Education." Wynberg, South Africa: Centre for Higher Education Transformation.

Staritz, Cornelia. 2010. "Competing in the Post-Quota and Post-Crisis World: Global Buyers and the Clothing Sector in Low-Income Countries" (Draft). World Bank, Washington, DC.

Statistics South Africa. 2010. "Quarterly Labour Force Survey, Sept. 2010." http://www.statssa.gov.za/qlfs/index.asp.

Stumpf, Rolf. 2010. "Higher Education Sub-Sector Analysis." Unpublished paper commissioned by the World Bank, Pretoria.

UNECA (United Nations Economic Commission for Africa). 2010. *A Technological Resurgence? Africa in the Global Flows of Technology*. Addis Ababa, Ethiopia: United Nations.

U.S. Department of Commerce. 1967. *Technological Innovation: Its Environment and Management*. Washington, DC: U.S. Government Printing Office.

Watkins, Alfred, and Michael Ehst, eds. 2008. *Science, Technology, and Innovation: Capacity Building for Sustainable Growth and Poverty Reduction*. Washington, DC: World Bank.

WEF (World Economic Forum). 2009. *The Africa Competitiveness Report: 2009*. Geneva, Switzerland: WEF.

World Bank. 2005. *Bangladesh: Growth and Export Competitiveness*. Report No. 31394. Washington, DC: World Bank, Poverty Reduction and Economic Management Sector Unit, South Asia Region.

———. 2010. Innovation Policy: *A Guide for Developing Countries*. Washington, DC: World Bank.

———. 2011a. "Closing the Skills and Technology Gaps in South Africa" (Draft, May). World Bank, Washington, DC.

———. 2011b. "Skills and Technology Absorption in Mauritius" (Draft, May). World Bank, Washington, DC.

Yusuf, Shahid, and Kaoru Nabeshima. 2011. *Some Small Countries Do It Better: Rapid Growth and Its Causes in Singapore, Finland and Ireland*. Washington, DC: World Bank.